GEOGRAPHY

THE BEHAVIORAL AND SOCIAL SCIENCES SURVEY
Geography Panel

Edward J. Taaffe, *Chairman*
The Ohio State University

Ian Burton
University of Toronto

Norton Ginsburg
University of Chicago

Peter R. Gould
Pennsylvania State University

Fred Lukermann
University of Minnesota

Philip L. Wagner
Simon Fraser University

GEOGRAPHY

Edited by
Edward J. Taaffe

Prentice-Hall, Inc., *Englewood Cliffs, N.J.*

 C–13-351353-X; P–13-351346-7. *Library of Congress Catalog Card Number 73–104854.* Printed in the United States of America.

Current printing (last number):
10 9 8 7 6 5 4 3 2 1

Prentice-Hall International, Inc. (*London*)
Prentice-Hall of Australia, Pty. Ltd. (*Sydney*)
Prentice-Hall of Canada, Ltd. (*Toronto*)
Prentice-Hall of India Private Limited (*New Delhi*)
Prentice-Hall of Japan, Inc. (*Tokyo*)

FOREWORD

This book is one of a series prepared in connection with the Survey of the Behavioral and Social Sciences conducted between 1967 and 1969 under the auspices of the Committee on Science and Public Policy of the National Academy of Sciences and the Problems and Policy Committee of the Social Science Research Council.

The Survey provides a comprehensive review and appraisal of these rapidly expanding fields of knowledge, and constitutes a basis for an informed, effective national policy to strengthen and develop these fields even further.

The reports in the Survey, each the work of a panel of scholars, include studies of anthropology, economics, geography, history as a social science, political science, psychology, psychiatry as a behavioral science, sociology, and the social science aspects of statistics, mathematics and computation. A general volume, *The Behavioral and Social Sciences: Outlook and Needs* (Englewood Cliffs, N.J.: Prentice-Hall, 1969), discusses relations among the disciplines, broad questions of utilization of the social sciences by society, and makes specific recommendations for public and university policy.

While close communication among all concerned has been the rule, the individual panel reports are the responsibility of the panels producing them. They have not been formally reviewed or approved by the Central Planning Committee or by the sponsoring organizations. They were reviewed at an earlier stage by representatives of the National Academy of Sciences and the Social Science Research Council.

Much of the data on the behavioral and social sciences in universities used in these reports comes from a 1968 questionnaire survey, conducted by the Survey Committee, of universities offering the PhD in one of these fields. Questionnaires were filled out by PhD-granting departments (referred to as the Departmental Questionnaire); by selected professional schools (referred to as the Professional School Questionnaire); by computation centers (referred to as the Computation Center Questionnaire); by university financial offices (referred to as the Administration Questionnaire); and by research institutes, centers, laboratories and museums engaged in research in the behavioral and social sciences (referred to as the Institute Questionnaire). Further information concerning this questionnaire survey is provided in the appendix to the general report of the Central Planning Committee, mentioned above.

Also included in the appendix of the report of Central Planning Committee is a discussion of the method of degree projection used in these reports, as well as some alternative methods.

THE BEHAVIORAL AND SOCIAL SCIENCES SURVEY COMMITTEE CENTRAL PLANNING COMMITTEE

Ernest R. Hilgard, *Stanford University*, Chairman
Henry W. Riecken, *Social Science Research Council*, Co-Chairman
Kenneth E. Clark, *University of Rochester*
James A. Davis, *Dartmouth College*
Fred R. Eggan, *The University of Chicago*
Heinz Eulau, *Stanford University*
Charles A. Ferguson, *Stanford University*
John L. Fischer, *Tulane University of Louisiana*
David A. Hamburg, *Stanford University*
Carl Kaysen, *Institute for Advanced Study*
William H. Kruskal, *The University of Chicago*
David S. Landes, *Harvard University*
James G. March, *University of California, Irvine*
George A. Miller, *The Rockefeller University*
Carl Pfaffmann, *The Rockefeller University*
Neil J. Smelser, *University of California, Berkeley*
Allan H. Smith, *Washington State University*
Robert M. Solow, *Massachusetts Institute of Technology*
Edward J. Taaffe, *The Ohio State University*
Charles Tilly, *The University of Michigan*
Stephen Viederman, *National Academy of Sciences*, Executive Officer

CONTENTS

Foreword v
Preface 1

1 Geography As a Social Science 5

Introduction 5
Illustrative Studies 8
Spatial Distributions and Interrelationships 8
Circulation 13
Regionalization 18
Central-place Systems 24
Diffusion 27
Environmental Perception 31

2 Methods and Achievements of the Field 37

Methods of Investigation 37
Cartographic Analysis 37
Mathematical and Statistical Techniques 44
Field Methods and Remote Sensing 49
Selected Research Directions 52
Locational Analysis 53
Cultural Geography 64
Urban Studies 72
Environmental and Spatial Behavior 89
Geography and Public Policy 99

3 Status, Trends, and Needs of Geography **104**

The United States and the World 104
Manpower 106
  Faculty and Enrollment 107
  Degrees and Graduate Enrollment 109
  Manpower Shortage 112
Research 114
  Research Trends 114
  Sources 116
  Foreign-Area Research 118
  Data Needs 119
Research Training 121
  Graduate Student Support 121
  Length of Training 124
  Mathematical Training 124
  Research Training Expenditures 127

4 Recommendations **131**

Appendix A
Geography Departments Participating in the Questionnaire Survey 137

Appendix B
Acknowledgments 139

Suggested Readings **141**

ILLUSTRATIONS

1 Land-Values Map 8
2 Land-Values Graph 9
3 Land-Values Diagram 10
4 Changing Population-Density Gradients 11
5 Chicago Land Values: 1910 and 1960 12
6 Population-Density Gradients—Indian Cities 13
7 Accessibility in Southeastern United States: Highway and Rail 15
8 Components of the Argentine Airline Network 16
9 Simulated Development of a Transportation Network 17
10 World Economic Development 18
11 Economic Development: New York State 19
12 Hospital Tributary Areas: Rail and Road 20
13 Commuting Zones: Chicago 21
14 Commuting Zones—Major U.S. Cities 22
15 Mormon Culture Region 23
16 A Hierarchy of Central Place Functions 25
17 Upper Midwest Trade Areas 26
18 Proposed Administrative Centers—Ghana 28
19 Diffusion of an Innovation 29
20 Simulated and Actual Diffusion 30
21 Expansion of Negro Ghetto: Actual and Simulated 32
22 Perceived Wilderness Areas 33
23 Perception of Flood Hazard 34
24 Residential Desirability 35
25 Population Dot and Density Map 39
26 Logarithmic Transformation 40

27 Population-Based World Map 41
28 Computerized Maps 43
29 Trend-Surface Diagrams 46
30 Settlement Trend Surfaces—Pennsylvania 47
31 Isodapanes: Mexican Iron and Steel 56
32 Transport Cost: U.S. Market 57
33 Theoretical Settlement Patterns 57
34 Optimal Flow Patterns: Wheat and Flour 58
35 Location Rent 59
36 Departure of Actual from Optimal Agricultural Production Patterns 60
37 Behavioral Matrix 61
38 Regional Simulation 63
39 Religious Pilgrimages: India 67
40 Settlement of the Great Columbia Plain 71
41 A Hierarchical Hinterland System 75
42 Graphs of City Size and Hinterlands 76
43 Migration-Sheds and Commutation Ranges: Upper Midwest 77
44 Air-Passenger Dominance 79
45 Development Profiles 80
46 Patterns of Economic, Family, and Ethnic Status 81
47 Commercial Centers and Tributary Areas: Chicago 82
48 Classification of Urban Commercial Areas 84
49 Megalopolis 85
50 Expansion of Minneapolis 85
51 Changing Commercial Patterns: Chicago, 1948–1958 86
52 Cultural Differences in Shopping Patterns—Mennonite and Modern Canadian 88
53 Comparative Hazard Perception: Flood Plain and Coastal Dwellers 92
54 Attendance at Recreational Reservoirs 95
55 Differing Space Preferences by Users 96

GEOGRAPHY

PREFACE

This report is designed to present a view of the field of geography in the United States in the late 1960s. Modern geography is a rapidly changing field, and the Geography Panel has attempted to reflect some of the diversity of activity characterizing contemporary geographic research. Stress will be laid on geography as a behavioral and social science and on the emerging public policy implications of geographic work.

This report is not designed to be comprehensive. It is selective on at least three counts: (1) it is concerned only with geography as a behavioral and social science, although geography has an important physical science component; (2) it is concerned with research rather than educational trends, although some attention is given to graduate training in the last chapter; (3) it does not attempt to cover all or even most research activity. Only four subfields are discussed, and, in order to provide reasonably clear illustrative studies, it was necessary to be extremely selective within each of these.

The first chapter, "Geography as a Social Science," provides a brief introduction and a series of illustrative studies designed to give readers unfamiliar with the field a brief overview of some of the types of research activities carried on by geographers today. Chapter 2, "Methods and Achievements of the Field," is divided into three parts: Investigative Methods, Selected Research Directions, and Geography and Public Policy. The methods discussion describes trends in the geographer's use of cartographic and mathematical analysis as well as his use of field and remote sensing techniques.

The discussion of research directions describes trends in four subfields: locational analysis, urban study, cultural geography, and environmental perception. The public policy discussion summarizes some of the geographer's involvement in work which has direct relation to public policy. The final chapter, "Status, Trends, and Needs of Geography," is a discussion of manpower, research, and research training in United States geography. This section includes a number of tables based largely on questionnaire data prepared by the Behavioral and Social Science Survey.

The interested reader might refer to previous surveys of geography for more inclusive treatments of the field. *Perspective on the Nature of Geography,* by Richard Hartshorne, provides an excellent summary of methodological debate in the field up to the mid-1950s. Relations between regional geography and other subfields are given particularly thoughtful attention in this work. The *Science of Geography,* a report issued in 1965 by an ad hoc Committee on Geography of the National Academy of Science–National Research Council, Earth Science Division, is the most recent survey. It stresses the role of geography in the study of the broad man-environment system from the point of view of spatial distributions and spatial relations. In addition to the surveys of cultural geography and locational analysis discussed in this report, *Science of Geography* includes a discussion of physical geography and of political geography.

The members of the Geography Panel would like to express their particular gratitude to Miss Jean King, Ohio State University, for her diligent and efficient organization of the typing and secretarial work, and to John Looman and Albert J. Ogren of Ohio State University for their cartographic assistance in redrawing the maps used in the report. Thanks are due the following members of the review panel for their thoughtful and helpful critiques of an early draft: Professor John C. Maxwell, Department of Geology, Princeton University; Professor Brian J. L. Berry, Department of Geography, University of Chicago; Professor Edward Ullman, Department of Geography, University of Washington; and Professor Beatrice B. Whiting, Department of Anthropology, Harvard University. Much help was provided by Warren Nystrom, Executive Secretary of the Association of American Geographers, and Walter Bailey, Earth Sciences Division, National Research Council, in advice and in the preparation of special survey material. A number of geographers cooperated with the panel by providing position papers and critiques of drafts of the report. The Committee on Geography of the Earth Sciences Division of the National Research Council, chaired by Saul B. Cohen, provided a number of helpful comments as did many of the chairmen of the PhD-granting departments of geography. We are particularly grateful to the geographers listed in Appendix B, who prepared brief position papers before the report was written. Finally, the members of the Geography Panel would like to acknowledge their indebtedness to Ernest R. Hilgard, Henry W. Riecken, and Stephen Viederman for their patience, critical comments, and constant guidance, which proved vital to the completion of the report.

1 GEOGRAPHY AS A SOCIAL SCIENCE

INTRODUCTION

Man's organization of his surroundings has been marked by rapid and accelerating change during the past hundred years. The changes in resource technology, scientific knowledge, and social institutions can no longer be measured in centuries, generations, or even decades, but only in shortening years. The growth of the social sciences as the fields of knowledge most concerned with the study of human behavior and institutions within that world of change is an accompanying fact. Geographers, like other social scientists, have participated in the growing pains as well as the growing achievements of the present era of scholarship.

A traditionally held view—that geography is concerned with giving man an orderly description of his world—makes clear the challenge faced by contemporary geographers. The world the geographer attempts to describe and interpret is enormously complex, and he faces great difficulty in choosing among diverse phenomena for his interpretation. He must not only consider a world of social, economic, and political behavior, but also a world in which such behavior is intertwined with the humane arts and the physical environment of man. In that world of bewildering diversity and constant change, geography has, at an increasing pace, reevaluated its inquiries and its theories.

Thus the history of geographic methodology has been long and complex, with overlapping generations of scholars bringing different perspectives to bear on what is significant for research. The contemporary stress is on geography as the study of spatial organization expressed as patterns and processes. This theme, together with several others, had its antecedents in nineteenth-century European geography. Among the major research themes developed at that time were: ecological studies of man-environment interrelations; studies of rapidly changing cultural landscapes, emphasizing syntheses of diverse but interrelated phenomena; and locational studies that emphasized geometrics of movement, size, shape, and distance. All these are represented in the theme of spatial organization, expressed as both pattern and process. Geographic study of the spatial organization of any area necessarily considers man-environment relationships and cultural landscapes. Such study is clearly integrative, and areas may be viewed as complexes of interrelated distributional patterns, lines of movement, and spatial processes, all involving change through time. Recent research emphases in geography also include the use of mathematical and statistical models in describing, analyzing, and understanding varied spatial patterns and processes; and an increasing concern with behavior in space, growing out of studies of the cultural perception and ecological interpretation of environment.

In the United States, geographic research has progressed relatively steadily with the other social sciences since the formation of the first graduate department at the University of Chicago shortly after the turn of the century. A typical pattern in American universities has been the development of geography within other departments—especially geology and, in some instances, history and anthropology—followed by the establishment of separate geography departments. In the middle thirties, the crisis of the Depression increased the involvement of geographers in national planning and research, especially in resource development. In the late 1930s and after World War II, there was a further surge of public interest in geography, related, in part, to the nation's increased international commitments. Entering graduate students during this period came to have a stronger systematic social science background than the prewar group, with its emphasis on geologic process and historical interpretation. Closer relations with the social sciences developed in some universities when area-studies programs were initiated and later, when interest heightened, in urban and resource development study. The growth of mathematical and theoretical work in the 1950s added impetus to these trends, and interrelations with economics and sociology began to develop more rapidly. Stimulus was also provided by the involvement of geographers in policy-oriented research into highway development, urban renewal, and resource management. Quantitative studies and locational analysis, which were initially stressed at only a few U.S. universities, developed rapidly with major inputs from Scandinavia. These trends have generated an increasing emphasis on geography as a research discipline. A concern with problem solving and more conscious attempts to develop theoretical structures have characterized this emphasis, which has been superimposed on the long-standing concern of geographers with their role in general education and teacher training.

Despite the relationships of geography with other social sciences in the United States and its increased attention to research on problems common to the other social sciences, some uncertainty remains, both among interested people in general and in other social and earth science disciplines, about the research directions of geography and its disciplinary role in the social sciences. Persistent misconceptions about geography stem from early school experience, in which the subject is often marked by detailed inventories of different parts of the world. Other lay views of the field are affected by the historic tie with geology, and many still regard geographers either as strictly physical scientists or as concerned exclusively with the relation between man and his physical environment.

The modern geographer's concern with spatial organization may be illustrated by a few brief examples of geographic studies made during the past 10 or 15 years. These simplified sketches are designed to survey a selected set of research themes rather than to provide a definitive summary of current work in each field. Although they cover only a small part of the discipline, they should be helpful as case examples of contemporary research questions with significant implications for public policy. Chapter 2, "Methods and Achievements of the Field," provides brief summaries of recent work in certain broad subfields which include, in varying proportions, all the research themes discussed below.

As in the other social sciences, there is an observable trend away from questions concerned primarily with detailing pattern, form, and structure, and toward the observation of behavior and analysis of the processes underlying structure. The geographer still observes, measures, and describes patterns found on maps, but he has become increasingly interested in the behavior or processes that formed those patterns. The sketches that follow have been selected to illustrate the interweaving of pattern and process in geographic study.

ILLUSTRATIVE STUDIES

Spatial Distributions and Interrelationships

A significant aspect of scientific inquiry is the search for order or pattern. Geographers have long been concerned with

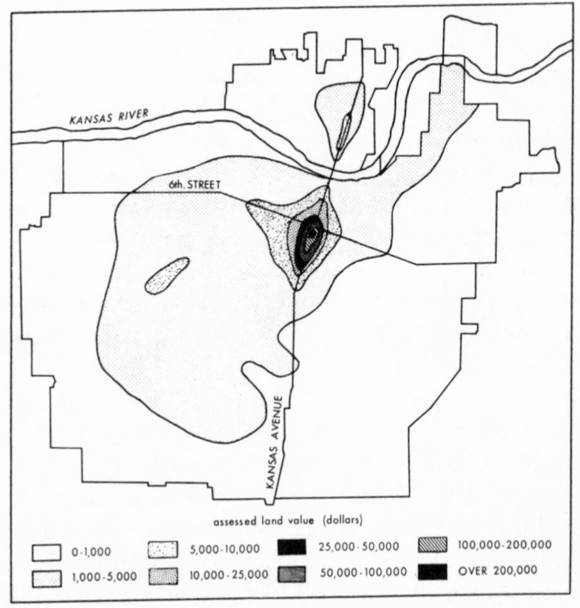

FIGURE 1 LAND-VALUES MAP

Assessed land values, Topeka, Kansas, 1954–59. From Duane S. Knos, *The Distribution of Land Values in Topeka, Kansas* (Lawrence: University of Kansas, Bureau of Business and Economic Research, May, 1962), Fig. 1.

providing accurate descriptions and explanations of the spatial patterns formed by the work of man on the face of the earth. In both regional and topical studies, considerable emphasis has been placed upon map analysis, both to portray complex patterns in a simplified fashion and to examine the manner in which they coincide with each other. Earlier work in agricultural geography examined relationships between the patterns shown on agricultural land-use maps and such physical patterns as landforms and soil. Later studies greatly expanded the number and nature of the variables considered. Patterns of settlement, transportation, voting behavior, social attitudes, and a variety of other measures of human activity were examined and compared at different geographic scales.

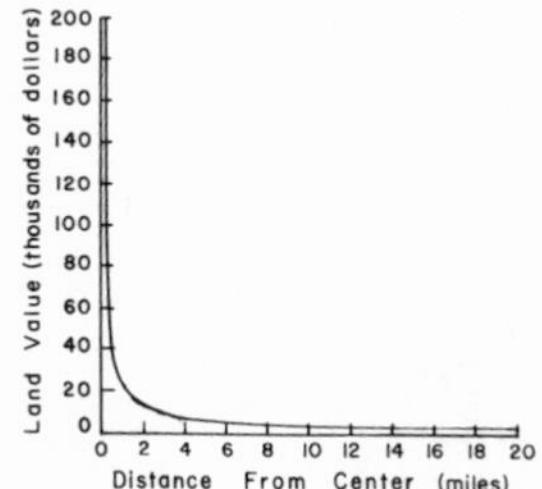

FIGURE 2 LAND-VALUES GRAPH

From Duane S. Knos, *The Distribution of Land Values in Topeka, Kansas* (Lawrence: University of Kansas, Bureau of Business and Economic Research, May, 1962), Fig. 3.

An example of the study of spatial patterns is the urban geographer's work on the interrelation between land values, population densities, and distance from the center of the city. Maps of land values have consistently shown a decline with distance from the central business district. Figure 1 shows a typical decline of land values in relation to distance from a central business district in Topeka, Kansas. The extraordinary steepness of this decline, which is evident in the graph (Figure 2) and in the three-dimensional representation (Figure 3), illustrates the high value placed on central locations in American cities. As in many other metropolitan areas of the United States, the land value surface can be approximated by postulating that values decline exponentially with distance from the central business district. Closer examination of the map also reveals other significant relationships, and more general descriptive equations were developed to express not only the relation between

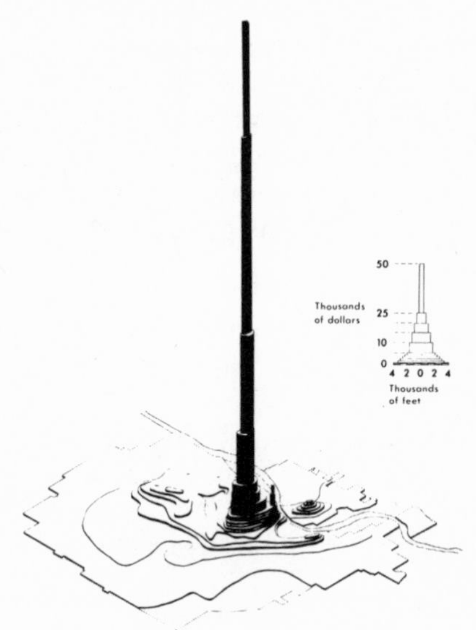

FIGURE 3 LAND-VALUES DIAGRAM

Three-dimensional representation of Topeka land values. This diagram shows both the extremely high values in the center of the city and the distribution of land values in the rest of the city.

From Duane S. Knos, *The Distribution of Land Values in Topeka, Kansas* (Lawrence: University of Kansas, Bureau of Business and Economic Research, May, 1962), Fig. 2.

land values and distance, but other important variables such as proximity to major thoroughfares, proximity to smaller secondary nodes of commercial activity in the city, and location within differing socioeconomic areas of the city.

The variation of such relationships with time was illustrated by studies of the changes in land value surfaces in Chicago. In 1910 there was a close relationship between land values and distance from the central business district, as well as from the elevated railway lines. In succeeding decades both land-value and population-density slopes tended to flatten out. Figure 4 shows the regular

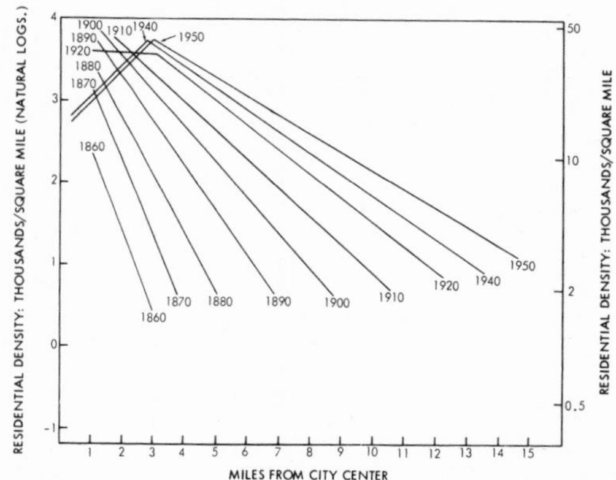

FIGURE 4 CHANGING POPULATION-DENSITY GRADIENTS

Graphic representation of the tendency for population density to decline exponentially with distance from the center of the city. As the city of Chicago has expanded, the gradient or rate of change in population density from the center of the city to the edge of the city has become steadily less.

From Brian J. L. Berry, *Geography of Market Centers and Retail Distribution* (Englewood Cliffs, N.J.: Prentice-Hall, Inc., 1967), Fig. 6.5.

manner in which population-density slopes flattened out in Chicago. A comparison of maps for 1910 and 1960 (Figures 5-A and 5-B) shows that, in 1960, there was no longer a consistent decline in land values in all directions from the central business district. On the 1960 map, the shaded areas representing low land values are concentrated in the western and southern parts of the city, and show the increasingly strong influence of such factors as racial composition and the provision of urban amenities. These factors, in turn, had different effects at different distances as well as in the different radial sectors leading from the center of the city.

Studies of distance-decay functions of land values and population densities are currently being expanded in an attempt to formulate more precisely their complex interrelationships with other factors.

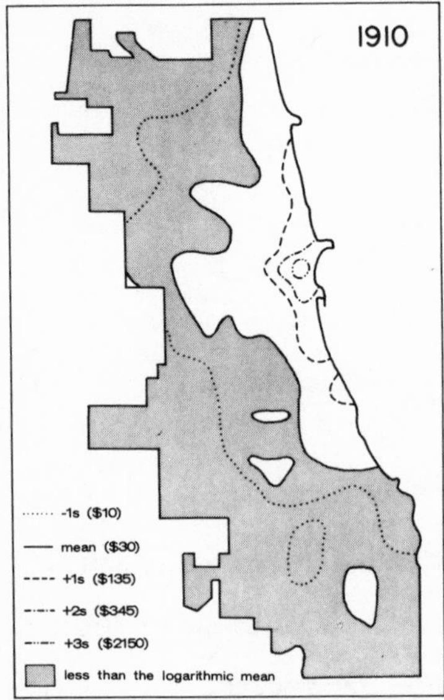

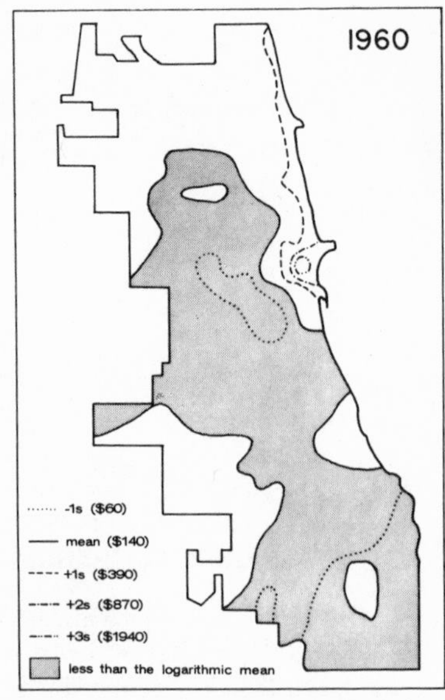

FIGURE 5 CHICAGO LAND VALUES: 1910 AND 1960

In 1910 there was a relatively regular decline of land values from the center of the city. The shaded areas, representing low land values, cover most of the outlying portions of the city. In 1960, the decline was much less regular. The concentration of shaded areas in the western and southern parts of the city indicates that low land values are associated with factors other than distance from the central business district.

After Maurice Yeates, "Some Factors Affecting the Spatial Distribution of Chicago Land Values, 1910–1960," *Economic Geography*, XLI, No. 1 (1965), 57–70, Figs. 1 and 2.

They are also being incorporated into cross-cultural investigations, where still further complications are being discovered. Figure 6, for example, shows population-density gradients for Indian cities which reflect such phenomena as dual gradients in cities with strong colonial origins (Group III) or more dispersed populations in some of the new, planned cities (Group IV).

As the relation between distance, land values, population density, and other features becomes clearer, it should be possible to isolate more precisely the effects of certain changes. For example, the impact a new expressway will have on population, land values, and the resulting tax base can be anticipated more effectively. The negative effects of atmospheric pollution become clearer as one compares maps of areas most affected by pollution with maps that show

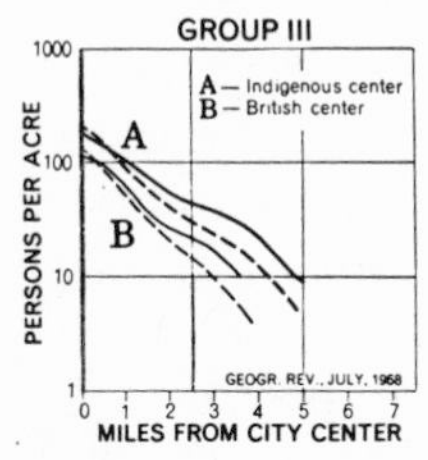

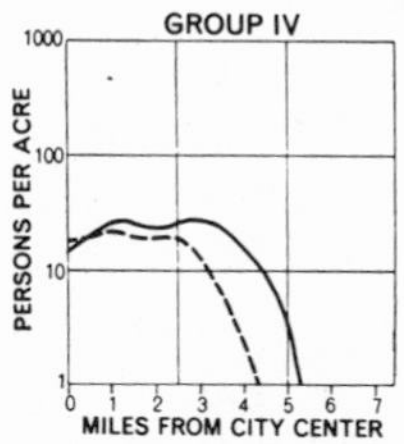

FIGURE 6 POPULATION-DENSITY GRADIENTS—INDIAN CITIES

Population-density gradients in Indian cities. The solid lines represent 1961 gradients. The dashed lines represent earlier gradients. The dual gradients in the Group III cities reflect their strong colonial origins. The new, planned cities that make up Group IV show a greater dispersion of population.

From John Brush, "Spatial Patterns of Population in Indian Cities," *Geographical Review*, LVIII, No. 3 (July, 1968), 362–92, Fig. 3.

where land values fall below the usual decline with distance, corrected for amenities and other factors. These effects, as well as those of changing population composition or new housing projects, form useful bases for formulating or modifying urban development policies.

Circulation

Patterns of movement or circulation also concern the geographer. Theoretical, as well as applied, studies of transportation systems, passenger and commodity-flow phenomena, and communications networks have been carried out. The blending of pure and applied geographic research in the analysis of networks is illustrated by a large number of studies directed toward defining and measuring the important and difficult concept of accessibility. For example, in a study of the accessibility of highway and railroad networks in the southeastern United States, a variety of mathematical models was employed to derive both overall measures of connectedness and individual measures of accessibility for each point in different types of networks. Overall measures of entire networks, derived from the ratios of actual circuits to the maximum number possible in the network, showed a generally higher level of accessibility in the proposed interstate highway system than in the existing rail network.

It is extremely difficult to tell from standard maps how the comparative accessibilities of individual cities such as Montgomery, Alabama, and Spartanburg, South Carolina, to all other cities differ in relation to rail and the highway networks. To evaluate individual city accessibility to all other cities in the network, the highway and rail systems were expressed as matrices of connections, and measurements were made of direct connections as well as multiple-step connections from each city to all other cities in the network. The resulting scale permitted a more precise ranking of cities in each network. As shown on Figure 7, Atlanta clearly ranks as the most accessible in both networks. Cities such as Richmond and Savannah, however, which occupy vital positions in the rail network are relatively less accessible in the network to be formed by the interstate highway system, whereas cities like Spartanburg and Chattanooga find their accessibility rankings markedly improved on this scale by the development of the highway network. The relative nature of the concept of accessibility is further emphasized if we consider the same centers in a network of airline linkages. Atlanta would still be the most accessible point, but larger centers such as Birmingham, Knoxville, and Norfolk are better connected to the other cities in the group by air than by surface transportation.

As more is learned about accessibility, it becomes possible to evaluate highway plans by their effects on accessibility patterns. A study in Ontario developed a means of comparing alternate choices for new highway construction by measuring the effects on the comparative accessibility of a group of cities. Other studies have related accessibility changes through time to differential rates of urban growth.

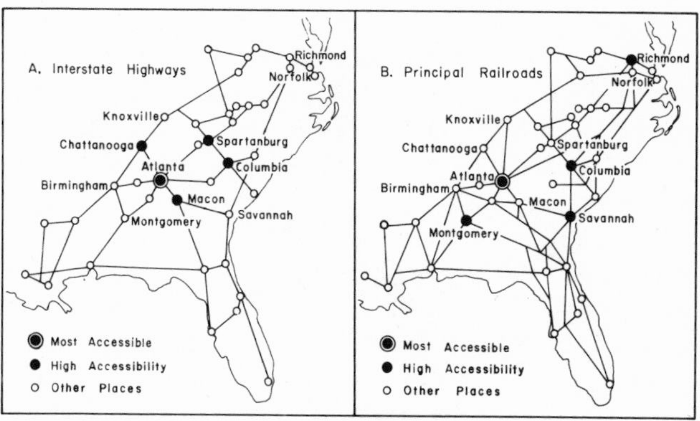

FIGURE 7 ACCESSIBILITY IN SOUTHEASTERN UNITED STATES: HIGHWAY AND RAIL

Comparison of the proposed interstate highway network and the existing railroad network in southeastern United States. The five cities with the highest accessibility ratings are named and indicated with black circles on each map. Atlanta is the most accessible city by either interstate highway or rail. Spartanburg and Macon, however, are more strategically located in the road network than in the rail network, whereas Savannah and Montgomery are more strategically located in the rail network.

After William L. Garrison, "Connectivity of the Interstate Highway System," *Papers and Proceedings of the Regional Science Association,* VI (1960), 121–37, Fig. 4.

Changes in a transportation network increase the accessibility of some cities relative to others. This may lead to increased rates of growth and expanded economies and tax bases for the favored cities. If greater control is to be exercised over the distribution of benefits from transportation network improvement, these predictive models can help planners to make more rational decisions. This is vitally important in areas where limited investment is available to develop a failing regional economy.

Another approach to the geographic study of transportation networks involves breaking down complex configurations into simpler, independent pieces. For example, in Argentina the air network has evolved over 40 years of changing demand and supply levels into a configuration that is spatially so complex that major patterns and

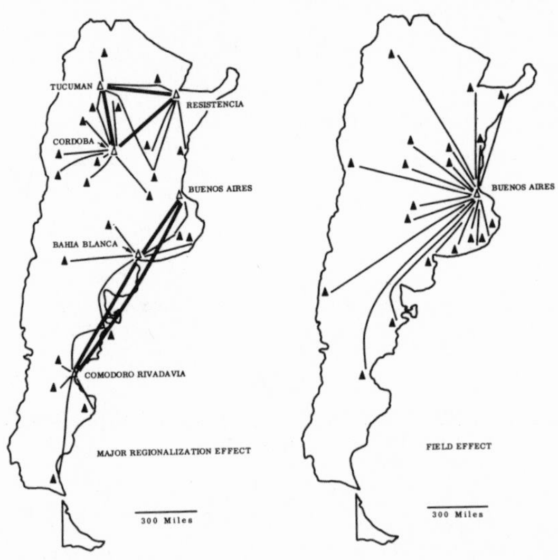

FIGURE 8 COMPONENTS OF THE ARGENTINE AIRLINE NETWORK

The complex Argentine airline network can be broken down by a combination of factor analysis and graph theory into maor regional effects, consisting of a coastal alignment and a triangular structure in the interior, and a field effect centered on Buenos Aires.

From William L. Garrison and Duane Marble, *The Structure of Transportation Networks,* TCREC Technical Report 62–11. Prepared by the Transportation Center at Northwestern University for the U.S. Army Transportation Research Command (Fort Eustis, Virginia: May, 1962), Figs. 15 and 16.

alignments are difficult to perceive. In Figure 8, an analysis of the air-network structure revealed that the complex total pattern could be decomposed into major regional effects centered on a coastal alignment of major cities and a triangular structure in the interior (map on left), together with a major urban field effect that centered routes on Buenos Aires (map on right).

Studies of transportation networks as spatial processes that unfold over time have been conducted. Simulations of transport network development have been based initially upon assumption of fixed urban and population distributions. Simple networks were "grown" from designated nodes in regular hexagonal lattices (Figure 9).

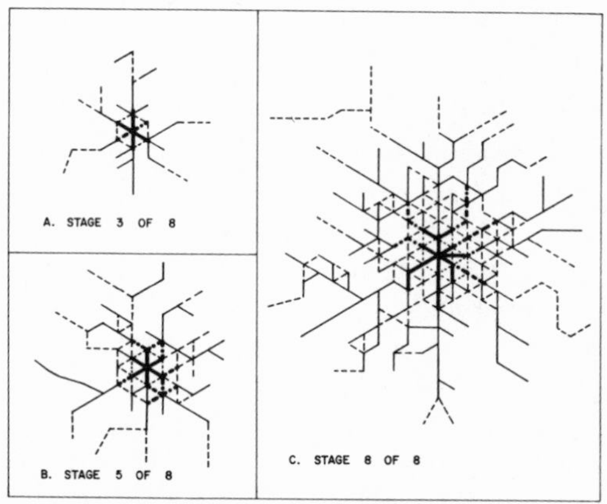

FIGURE 9 SIMULATED DEVELOPMENT OF A TRANSPORTATION NETWORK

Simulated growth of a transportation network based on probabilities derived from preexisting population and urban distribution, previous transportation development, and other factors.

From Richard L. Morrill, *Migration and the Growth of Human Settlement*, "Lund Studies in Geography, Series B, Human Geography, No. 24" (Lund: C. W. K. Gleerup, 1965), Figs. 7.10–7.17.

After insight had been gained through work with such abstract spatial processes, more realistic and ambitious models that took the actual historical experience of a region into account were devised and simulations of plausible patterns of road development together with accompanying changes in urban structure and patterns of rural-to-urban migration over an 80-year period were seen. As comprehensive models of this sort are developed, they appear to be unusually promising tools for urban and regional planning. Alternate highway plans, for example, might be evaluated not only for their efficiency but also for their predicted impact upon urban growth and population shifts.

Regionalization

Regionalization studies are studies of spatial structure that involve both patterns and linkages in complex interrelationships. Regionalization is, in part, a classificatory matter. More generally, however, it deals with the characteristics of all areas of organization that were developed by men of a particular society to achieve purposes that evolved as part of their culture.

Regionalization may be on the basis of relatively few dominant criteria that a scholar with an intimate knowledge of the area selected, as in many studies of culture regions. An area may also be regionalized on the basis of a large number of variables that show a certain degree of homogeneity within identifiable areas. When regionalization involves dealing with many criteria, the complexity may be reduced by grouping. In recent years, expanded computer facilities have greatly extended the geographer's ability to classify

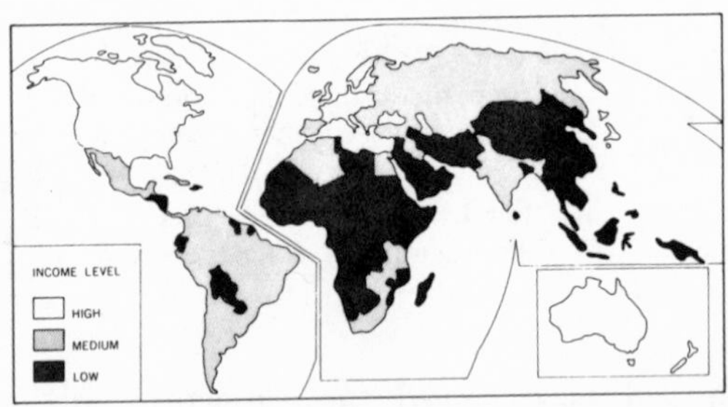

FIGURE 10 WORLD ECONOMIC DEVELOPMENT

Classification of economic development based on a factor analysis of 43 different economic and social indices.

After Norton Ginsburg, *Atlas of Economic Development* (Chicago: University of Chicago Press, 1961), p. 111.

areas on the basis of many criteria simultaneously, and objective classificatory procedures (widely shared with other social and behavioral sciences), are in constant use. Figure 10, an illustration of this process on the world scale, indicates varying degrees of economic development on a single, composite index, that weights 43 economic and social indices. The map displays the great regional disparities between the developed and underdeveloped portions of the world.

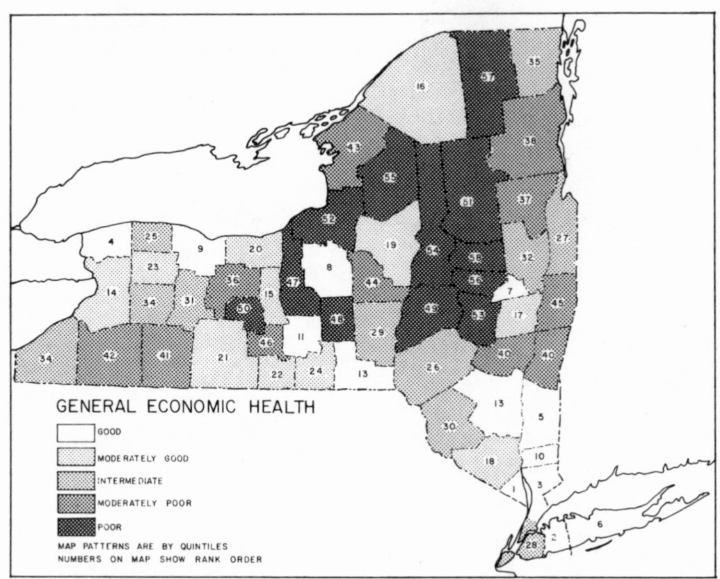

FIGURE 11 ECONOMIC DEVELOPMENT: NEW YORK STATE

Counties of New York State classified as economically "healthy" or "unhealthy," based on scales derived from a large number of socioeconomic variables. The counties have been ranked on these scales and divided into five categories of equal size.

After John Thompson, *et al.*, "Toward a Geography of Economic Health: The Case of New York State," *Annals*, Association of American Geographers, LII, No. 1 (1962), Fig. 12.

Smaller areas may be similarly classified. Figure 11 shows the counties of New York state classified—on scales derived from a large number of relevant variables—as economically "healthy" or "unhealthy." Groupings of any desired size at any geographic scale may be obtained from such procedures. The result will be the identification of economically and socially similar regions which will form a useful base for the allocation of development funds or the organization of administrative and planning districts.

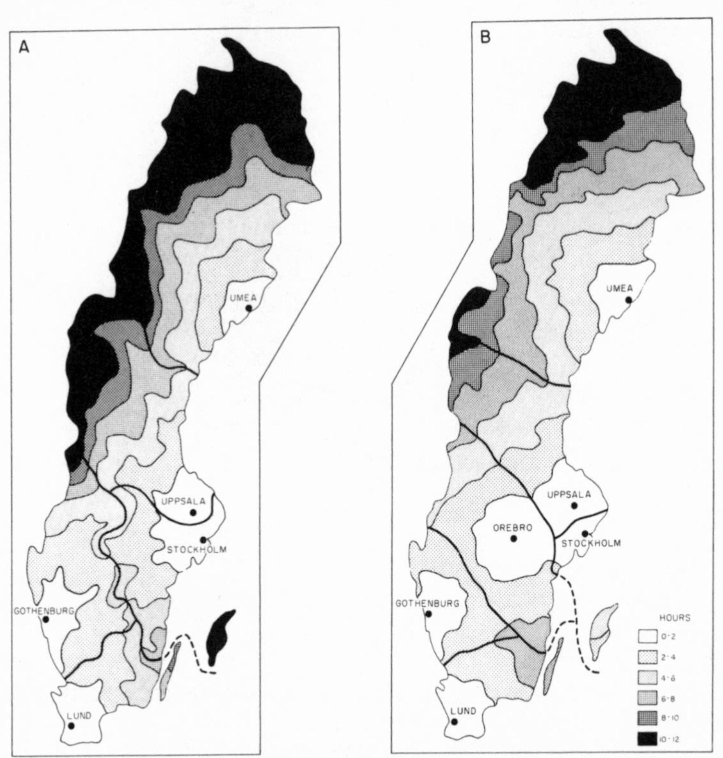

FIGURE 12 HOSPITAL TRIBUTARY AREAS: RAIL AND ROAD

The isochrones, or lines of equal time-distance, show travelling times to proposed regional hospitals by train, bus, and boat connections (A), and by passenger car (B). The heavy lines enclose presumed tributary areas for each hospital, based on minimum travel times.

After Sven Godlund, *Population, Regional Hospitals, Transport Facilities and Regions*, "Lund Studies in Geography, Series B, Human Geography, No. 21" (Lund: C. W. K. Gleerup, 1961), Figs. 4 and 8.

Regionalization may also be based on spatial interaction. Here the criteria for regionalization are primarily linkages or nodality rather than the identification of homogeneous areas. Figure 12 shows maps of tributary areas for regional hospitals, based on potential interaction, as indicated by hourly travel zones from each large regional hospital to all parts of its area by train and by passenger car. In a study of possible locations for new hospitals in Sweden, a

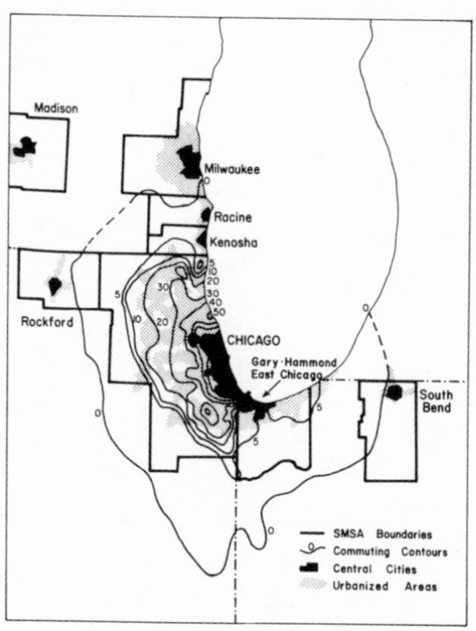

FIGURE 13 COMMUTING ZONES: CHICAGO

Contour lines enclose areas with indicated percentages of Chicago commuters. Shaded areas indicate urbanized areas.

From Philip Reese, *The Factorial Ecology of Metropolitan Chicago, 1960.* Unpublished M.A. thesis, University of Chicago, Department of Geography, 1968.

series of hourly travel-zone maps was prepared for each of several alternate locations. By superimposing each map on a detailed dot map of population, it was possible to determine the percentage of total population within one, two, three, or more hours of a regional hospital, and to obtain comparable figures for total transport costs for the alternate locations. Similar procedures were applied to other forms of transportation, effects of future hospital expansion, and population shifts. With localized data available, computerization of these procedures is quite feasible and the range of alternatives can readily be widened.

The metropolitan area is emerging as a type of region of considerable significance as a functional unit and base for planning. Recent studies by geographers for the U.S. Bureau of the Census considered problems involved in delimiting metropolitan areas. For example, Figure 13 shows the extensive zone within which there is some

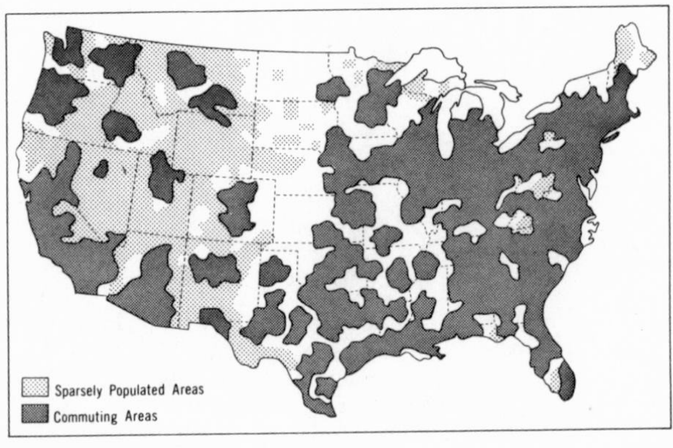

FIGURE 14 COMMUTING ZONES—MAJOR U.S. CITIES

The dark areas are those which sent commuters to the central city of a standard metropolitan statistical area (SMSA) in 1960. They included 87 percent of the population. The lighter shading represents essentially unpopulated areas (less than 1–2 persons per square mile). Thus, only the white areas lay beyond the zone of metropolitan influence in 1960.

After B. Berry and J. Meltzer, eds., *Goals for Urban America* (Englewood Cliffs, N.J.: Prentice-Hall, Inc., 1967), Fig. 1.

commuting to Chicago. These studies have resulted in the mapping of a set of commuting zones around major U.S. cities in order to provide a more realistic picture of the functional reach of U.S. cities beyond their political boundaries (Figure 14).

Regionalization may also be studied as a dynamic spatial process.

Figure 15 shows the Mormon culture region. The core area is a centralized zone of concentration and control; the domain includes those areas in which Mormon culture is dominant; the sphere is the

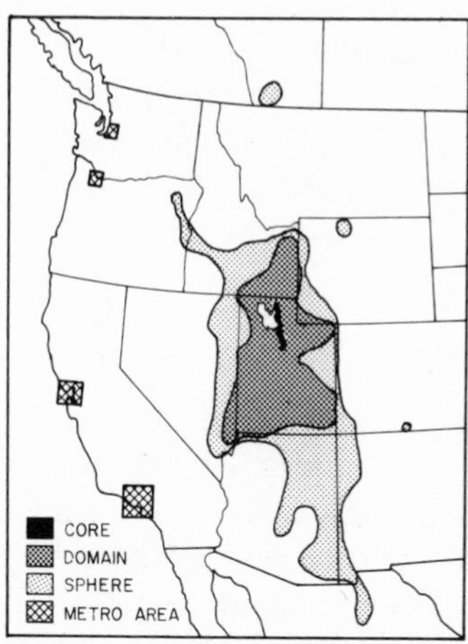

FIGURE 15 MORMAN CULTURE REGION

The black area represents the Wasatch Oasis, the centralized zone of concentration and the focus of organization of the Mormon culture that contains about 40% of the Mormon population. Another 28% lives in the area of dark shading, where the Mormon culture is dominant, but with markedly less intensity and complexity of development than in the core. The lightly shaded area, which represents the sphere of the Mormon culture, contains 13% of the total. This is a zone of outer influence and peripheral acculturation, in which the culture is represented only by certain of its elements or where its peoples reside as minorities. In the metropolitan areas represented on the map, Mormon minorities live in the indicated cities, and these groups constitute 17% of the total Mormon population.

After Donald W. Meinig, "The Mormon Culture Region: Strategies and Patterns in the Geography of the American West, 1847–1964," *Annals*, Association of American Geographers, LV, No. 2 (June, 1955), Fig. 7.

zone of outer influence where Mormons are often a minority. The region was formed by the diffusion of Mormon culture from the Salt Lake core to other parts of Western North America. The present shape of the region may be interpreted as a compromise between an expanding internal force, represented by Brigham Young's desire to colonize southward into Mexico as well as into Canada, and contracting external forces, represented by federal power and incursions by railroad builders and mining interests. Regions of this sort are undergoing constant change, and their development within both cities and regions epitomizes continuing tensions between the local-scale organization of space by a culturally cohesive group, and the national-scale trend toward increased standardization and focus on a few metropolitan centers.

Central-place Systems

Another approach to investigating spatial structure, similar to regionalization, is the study of central-place systems. Continued map analysis of settlement patterns has made it clear that there are certain regularities in the number and spacing of centers. There tends to be a nested hierarchy of centers, with a group of villages tributary to a town; a group of towns tributary to a city; a group of cities tributary to a larger metropolitan center. Characteristic clusterings of functions within villages, towns, and cities have been identified. Some examples of village-, town-, and city-level functions are shown on the accompanying maps of western Iowa. Figure 16-A shows the widely dispersed, short-distance travel characteristic of grocery shopping. Villages include these functions, as do towns and cities. The travel characteristic of women's clothing shopping (16-B) shows more concentration on the town-level centers. An example of a function found at a still higher level in the hierarchy of centers is shown in 16-C. The lines on this map represent newspaper orientation and show a concentration on Omaha-Council Bluffs to the west and Des Moines to the east. Figures 17-A and 17-B indicate that similar patterns exist further up the geographic scale. Metropolitan areas such as Fargo, Sioux Falls, and Duluth have large trade areas at the wholesale-retail level, each containing a constellation of cities, towns, and villages similar to the Iowa example. These trade areas, in turn, fall inside the great metropolitan field of the Twin Cities. All these trade areas are elongated in directions away from their nearest competitors.

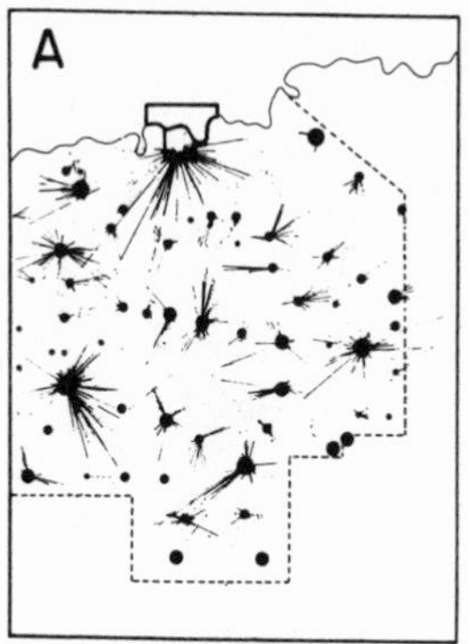

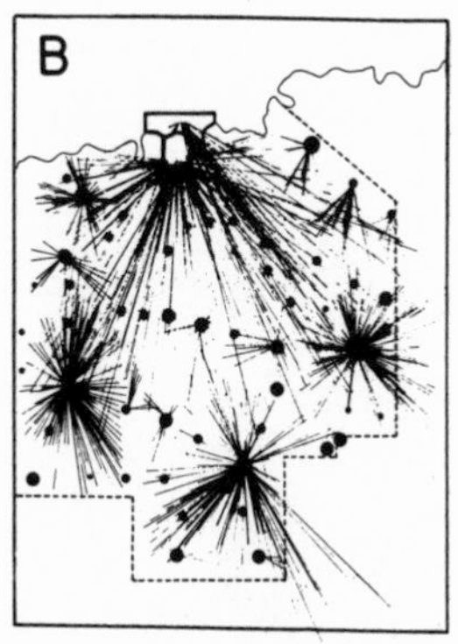

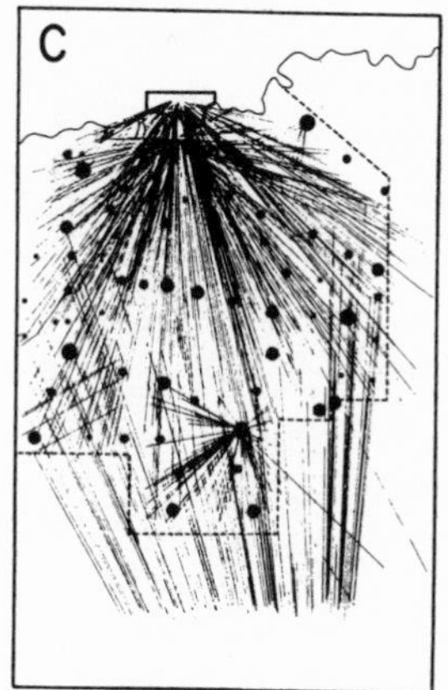

FIGURE 16 A HIERARCHY OF CENTRAL PLACE FUNCTIONS

Lines indicate shopping travel patterns for groceries (A), travel patterns for women's clothing (B), and newspaper orientation (C). The grocery shopping is characterized by short-distance trips to villages, towns, and cities; clothing is characterized by medium-distance travel to the towns and cities; newspaper orientation is focused on the larger centers of Omaha–Council Bluffs to the west and Des Moines to the east.

From Brian J. L. Berry, *Geography of Market Centers and Retail Distribution* (Englewood Cliffs, N.J.: Prentice-Hall, Inc., 1967), Figs. 1.9b, 1.9d, and 1.9f.

Thus, an area is organized by human activity and requirements into a system of functional regions nesting within each other, from the smallest hamlet to a major metropolitan focus. Studies have been made in the United States and other parts of the world that

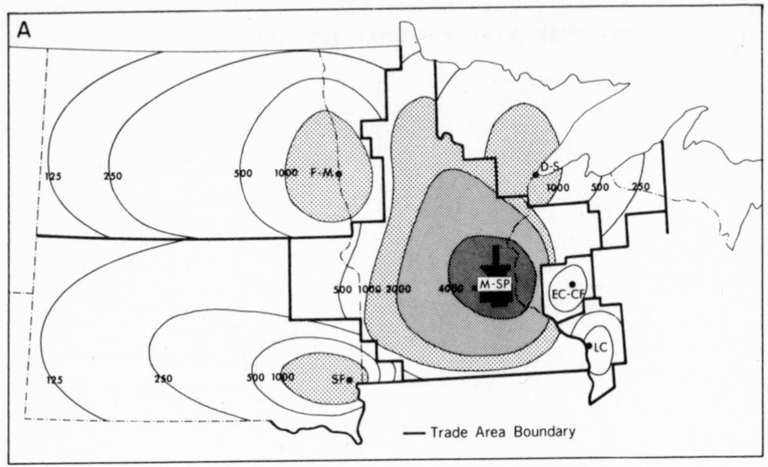

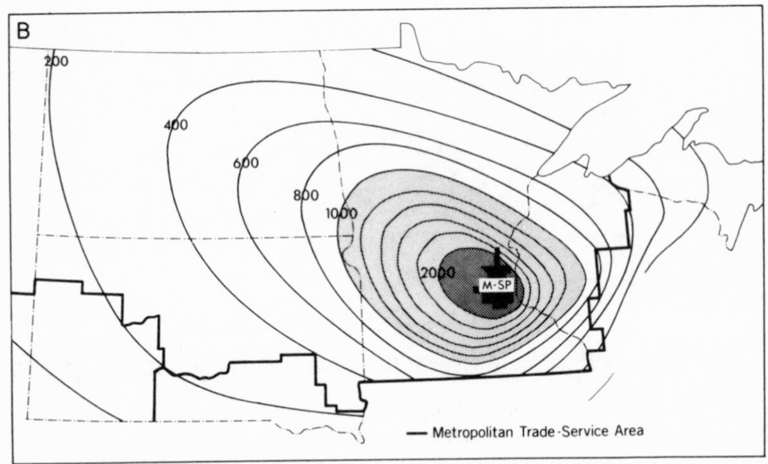

FIGURE 17 UPPER MIDWEST TRADE AREAS

The reach of the primary wholesale-retail centers in the upper Midwest is shown in A. Contour lines representing phone calls per capita from smaller centers were mapped to isolate trade area boundaries adjusted to correspond with counties. B represents the economic reach of Minneapolis–St. Paul and shows its metropolitan service-area boundaries relative to such centers as Des Moines, Omaha, Denver, Seattle, and Portland. Contour lines represent per-capita phone calls from the primary centers to the metropolis.

After John R. Borchert and Russell B. Adams, *Trade Centers and Trade Areas of the Upper Midwest*, "Urban Report No. 3, Upper Midwest Economic Study" (Minneapolis: University of Minnesota, 1963), Figs. 9 and 8.

examine relationships between four closely related characteristics of these urban systems: size of center, spacing between centers, functions found in centers, and tributary area of centers. These have been expressed relatively precisely and have been found to be reasonably similar and predictable in areas of given densities and socioeconomic characteristics.

Studies have indicated processes of change at work through time in these city systems. The influence of the large metropolitan centers has been steadily increasing. With better highways and increased mobility it has become easier for farmers and residents of smaller urban centers to take advantage of the many conveniences of the large centers, thus increasing market area sizes and forcing functions upward in the hierarchy, and affecting the fast-disappearing hamlets first, then the villages. A Saskatchewan study indicated that nearly half of the hamlets in a selected area had disappeared during the last 20 years and two-thirds of the villages had declined to hamlet status.

The central-place systems provide a basis for one approach to the development and location of government and social services of all kinds. In the reclaimed areas of the Zuider Zee in Holland and in the northern Negev of Israel, planned settlement of previously unoccupied areas has proceeded along lines involving central-place hierarchies. In Canada and in parts of the United States, central-place hierachies have influenced plans for relocating services and facilities because of declining rural populations and expanding metropolises. In developing countries such as Ghana, where towns are growing in size and significance, research into the geographic structure of the central-place system has disclosed major gaps in the administrative coverage of the country and has provided one possible logical base from which decisions may be made to rearrange the present pattern of administrative centers and supplement them with new regional offices if necessary (Figure 18).

Diffusion

Studies of spatial diffusion illustrate processes whereby patterns are developed through time. These studies have been characteristic of geographic work—particularly in cultural geography—

for many years, but they have been much influenced recently by the development of models for simulating the diffusion of innovation. An early study dealt with the adoption of an agricultural in-

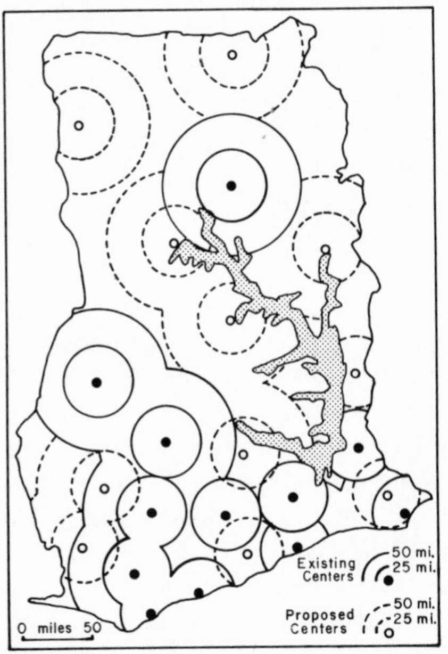

FIGURE 18 PROPOSED ADMINISTRATIVE CENTERS—GHANA

Present and proposed pattern of administrative centers in Ghana. After David Grove and Laszio Huszar, *The Towns of Ghana* (Accra: Ghana Universities Press, 1964).

novation in Sweden. Figure 19 shows the sequence of adoption of one of these innovations. There is a clear "neighborhood" effect as each set of adopters seems to have been particularly influenced by nearby innovators. The probability of adoption by one farmer thus seems to decline with his distance from some other farmer who has already accepted the innovation. A simple grid was therefore developed, based on the assumption of a set of declining probabilities surrounding each adopter. A simulation model was developed for an area in Sweden with decision-rules based on the effects of this probability grid as well as on other probabilities based on such things as the actual distribution of poulation and the effects of certain barriers such as mountains and lakes. This model provided an initial probability surface and a set of random numbers was then drawn to assign a certain number of adopters over the area. For each successive time period a new probability surface had to be

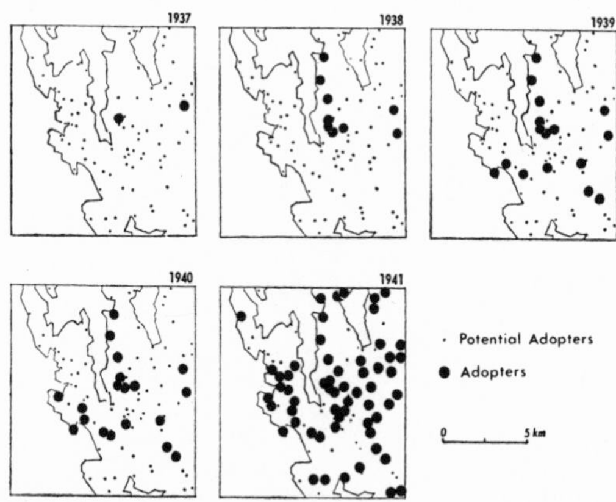

FIGURE 19 DIFFUSION OF AN INNOVATION

Spread of systematic control of bovine tuberculosis in Sweden. Small dots represent potential adopters; large dots represent actual adopters.

From Torsten Hagerstrand, "A Monte Carlo Approach to Diffusion," *European Journal of Sociology*, VI (1965), 43–67, Fig. 1B.

generated, because the distribution of adopters from the previous time period would affect the probability of the innovation being adopted at any given area. A series of simulated patterns was then run to assign adopters to the particular cells indicated on the probability surface for each time period in sequence. Figures 20-A and 20-B compare the results of a typical simulation with the observed spread of an innovation. The relationship is reasonably close, as it was in the other simulations, suggesting that the particular decision-

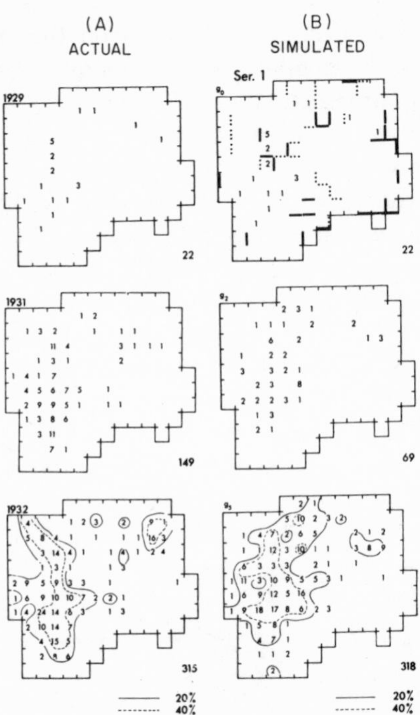

FIGURE 20 SIMULATED AND ACTUAL DIFFUSION

Figure 20A shows the actual spread of an agricultural innovation. The figures in the cells give the absolute number of adopters at each time. Potential adopters are relatively evenly distributed through the area. For 1932, isolines indicate the relationship between adopters and potential adopters. Figure 20B shows a simulated spread of an innovation through the same area. The lines on the first diagram represent barriers to communication similar to those in the actual area.

From Torsten Hagerstrand, "A Monte Carlo Approach to Diffusion," *European Journal of Sociology*, VI (1965), 43–67, Figs. 3 and 4.

rules employed provide a fair description of the process of spatial diffusion of that innovation. Closer examination of several simulations, however, revealed certain consistent or nonrandom departures from the actual pattern. These departures have been the basis for further studies of such things as the varying levels of adoption among different socioeconomic groups and the tendency for certain innovations to diffuse through an urban hierarchy. Studies of the diffusion of radio and television stations in the United States, for example, have shown a "colonizing" type of model with the largest centers adopting the innovation first, then spreading it to the smaller centers surrounding them.

Still another variation on the spatial diffusion process is in the study of intraurban migration. A simulation model much like the one used in the innovation study was applied to the expansion of the Negro ghetto in Seattle. The probabilities used, however, took account of the marked tendency for the probability of Negro migration to decline sharply with distance from the edge of the ghetto and for the probability of a migration into a given block to increase considerably as the first few Negro migrants enter. Figures 21-A through 21-D represent actual and simulated expansion in the decades between 1940 and 1960. In general, the decision-rules employed in the model seem to provide a reasonably good description of the process of ghetto expansion. Certain discrepancies such as those on the west side of the park suggest weaknesses in the model; current research is proceeding toward refinements that relate these models more closely to the actual residential mobility of both Negro and white residents at the active fringes of expanding ghetto areas.

As these models of spatial diffusion and migration ramify through empirical testing and reformulation, they provide an improved understanding of the processes underlying the spread of different ethnic groups through a city, or the way in which a potentially beneficial technological innovation is likely to spread through an economically depressed area. This, in turn, should permit better prediction of the consequences of alternative policies that concern such issues as the provision of new housing, the nature of public-information programs, or the location of demonstration farms in depressed agricultural areas.

Environmental Perception

Studies in environmental perception are concerned with efforts to understand how men structure in their own minds the world around them. Some studies give explicit attention to the ways in which men perceive elements of their natural environment and how they apprehend resources or natural hazards such as floods and droughts. Other studies treat man's views of landscapes, especially

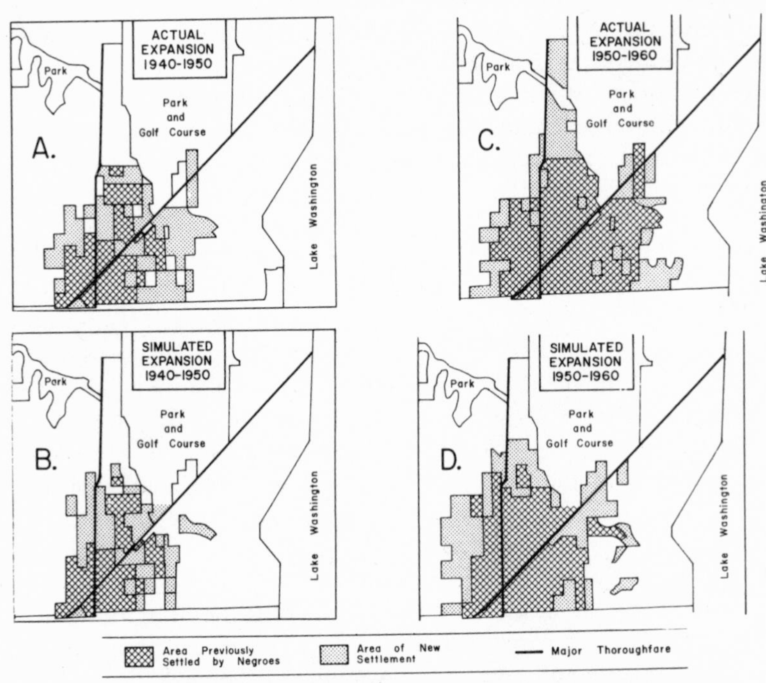

FIGURE 21 EXPANSION OF NEGRO GHETTO: ACTUAL AND SIMULATED

Figures 21-A and 21-B show the actual expansion of the Negro ghetto in Seattle in the 1940–1950 period as compared to the simulated expansion. Figures 21-C and 21-D show the actual and simulated expansion of the ghetto during the 1950–1960 period.

After Richard L. Morrill, "The Negro Ghetto: Problems and Alternatives," *Geographical Review*, LV (1965), 339–61, Figs. 9, 10, 12, and 13.

in urban areas, and his perceptions of differing spatial organizations and attitudes toward places, as shown in "mental maps."

One example of a perception study dealing with the natural environment is the investigation made into wilderness perception in the Quetico-Superior wilderness area of the U.S.–Canadian border. The purpose of the research was to study differences in wilderness perception between various classes of recreational users as well as of the officials responsible for the management of the area. It was found that the area considered to be wilderness varied consistently with the type of use. While the majority of auto and boat campers considered most of the area to be wilderness, the majority of the canoeists were more exacting in their standards and considered much less of the area to be wilderness (Figure 22).

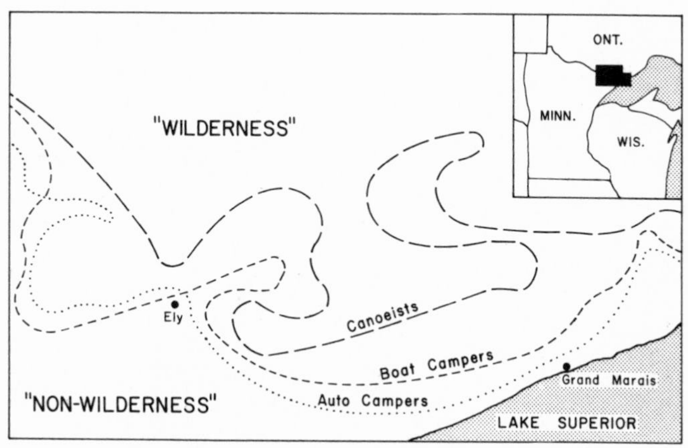

FIGURE 22 PERCEIVED WILDERNESS AREAS

Portions of the Quetico-Superior area perceived as wilderness by different groups of campers. Lines separate "wilderness" from "non-wilderness" as perceived by auto campers, boat campers, and canoeists. The canoeists were most exacting in their standards, and considered less of the area to be wilderness than did the other two groups.

After Robert C. Lucas, "Wilderness Perception and Use: The Example of the Boundary Waters Canoe Area," *Natural Resources Journal*, III, No. 3 (1964), 394–411, Fig. 3.

A more formal statement of the wilderness criteria for different types of users has been developed from this study. These criteria can be used as a basis for recommendations on zoning and visitor regulation, designed to preserve the wilderness characteristics deemed essential by different groups and to provide for more public use in the face of steadily increasing demand.

Another group of environmental-perception studies deals with human attitudes toward natural hazards. Figure 23 shows the varia-

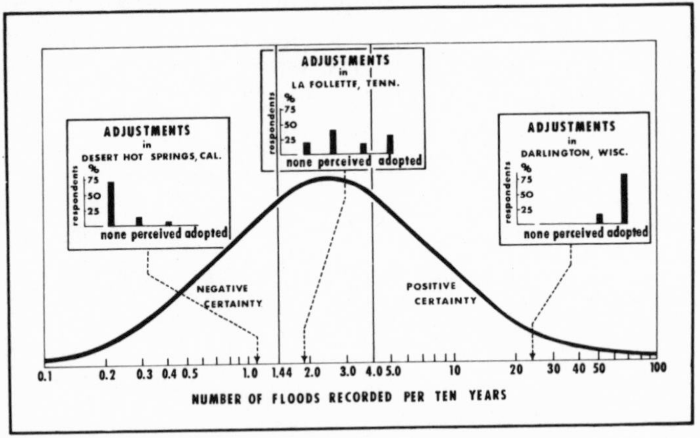

FIGURE 23 PERCEPTION OF FLOOD HAZARD

Diagram relating frequency of floods to perception and adjustment. Superimposed on a log-normal distribution of 496 urban places for which flood-frequency data were available are insets showing perception and adjustment at three sites. Adjustment is scaled from total ignorance through two levels of adjustment to adoption. Below a flood frequency of 1.44 per ten years, adjustment is virtually nonexistent; above a frequency of 4.0, some adjustment is almost universal.

After Robert W. Kates, *Hazard and Choice Perception in Flood Plain Management* (Chicago: University of Chicago, Department of Geography Research Paper No. 78, 1962), Fig. 9.

tions in the perception and adoption of adjustments designed to reduce flood damage in selected sites. With a low frequency of flood occurrence, there may be no adjustments. At a high frequency of occurrence (four or more in ten years) some adjustment is almost universal. In the intermediate frequency (approximately one and one-half to four floods in ten years) the adjustment behavior is more varied. These studies have made significant contributions to the development of more flexible and realistic policies by the federal government.

A third example of a geographic-perception study deals with "mental maps," which summarize the major viewpoints of a group of people when various areas are considered for their residential desirability (Figure 24). The view of the United States from Pennsylvania was based on a component analysis filtering out the most general viewpoint of the sample group. The perception surface shows a high ridge of desirability along the West Coast, where the south central coastal zone is preferred, and desirability declines in the surrounding areas. A "mental basin" (low desirability) appears in Utah and Nevada, but the perception surface rises again in Colorado. While the general trend is downward from west to east, the surface shifts 90 degrees to a north-south orientation in the middle of the country. With the exception of a Dakota low, the Deep South is the lowest trough on the entire map, and the preference contours

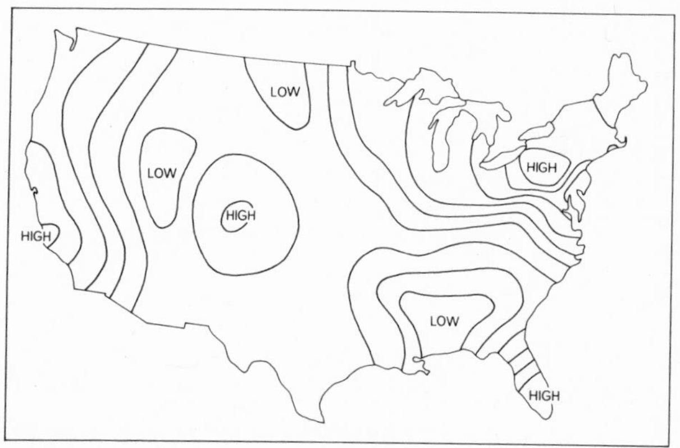

FIGURE 24 RESIDENTIAL DESIRABILITY

Mental map showing areas of high and low residential desirability as evaluated by a group of Pennsylvania college students.

After Peter Gould, "Structuring Information on Spacio-Temporal Preferences," *Journal of Regional Science,* VII, No. 2 (Supplement), (Winter, 1967), 259–74, Fig. 1.

rise northward, first slowly, then more rapidly to the highest zone in the home state, in this case Pennsylvania.

Studies of attitudes toward different areas both within and between countries have much potential use in research on migration and regional economic development. American geographers in Africa are currently doing a study in which perception surfaces will be used as the basis for designing a set of incentive wage differentials between regions to discourage the increasing concentration of the highly educated elite in just a few large cities. Perception studies are also under way in urban areas, where stress caused by the real or presumed hostility between neighboring groups is examined as a strong force behind much intraurban migration. Scales developed for tolerance levels in different urban environments may be related to individual tendencies to move or stay as stress develops in a particular neighborhood.

These sketches have provided a few glimpses of the diversity of contemporary geographic research. The geographer's concern with spatial organization involves him in problems of the city, regional development, and environmental control. In the city he studies changing population patterns, expanding ghettos, and increasing reach of the great metropolitan areas. In the region, he studies changing circulation and accessibility patterns; the rise and decline of large, medium, and small centers; and the spread of people and ideas from place to place. He examines man's own perception of his environment and suggests ways of organizing and preserving wilderness and other natural resources, as well as ways of coping with flood, drought, and other natural hazards.

2
METHODS AND ACHIEVEMENTS OF THE FIELD

METHODS OF INVESTIGATION

Because geographers bring a different viewpoint to the study of many of the same phenomena and problems treated by other social and behavioral scientists, it is of particular concern to survey briefly their methods of study. These include the map as an analytical tool, mathematical and statistical models in spatial problems, and field investigation and remote sensing. Since the emphasis in this section is on methods of investigation rather than on findings, the discussion here will have to be somewhat more technical than in other sections of the report.

Cartographic Analysis

Maps have been closely associated with the work of the geographer from the earliest days of exploration. In those days, maps were used to solve problems of moving from place to place, since they provided relatively accurate information on place location, stream navigability, and topography. Today, geographers can also use maps to deal with social and behavioral questions, both for communicating findings and for the analysis of problems. Map use for problem analysis has greatly increased in recent decades and its utility has been enhanced by current developments in mathematical and statistical analysis and the use of computers.

In the sense that they compress, abstract, and simplify reality, maps serve as models that retain the spatial relationships and juxtapositions relevant for particular purposes of analysis. As analytical tools maps possess at least four useful characteristics: (1) they are highly efficient for certain types of data storage; (2) they permit a variety of multidimensional measurement; (3) they may be used to transform surface characteristics in the plane; and (4) they may be used for testing hypotheses on spatial organization.

Maps are extremely efficient devices for the storage of spatially associated data in a form that allows instant associative recall. Although more information, together with locational coordinates, can be stored in computers, spatial associations between the bits of information cannot readily be viewed. The number of possible combinations of phenomena evident in a map that has a fairly small number of data variables plotted becomes quite sizable when one realizes that such basic spatial quantities as distance, direction, connectedness, contiguity, and area may now be associated with each set of variables. It has been estimated that the average U.S. topographic quadrangle—showing land elevations, roads, and certain settlement forms—contains over 100 million separate bits of information, more than the average map reader could absorb in a year's time, and the topographic sheets of many other countries contain even more. One example of the importance of this informational role of maps is the development of planning atlases. Atlases being produced at the state, national, and regional levels contain useful cartographic representations of detailed social, economic, and political data.

Second, maps serve as models that permit the measurement and analysis of both static and dynamic spatial relationships. On maps that show complex land-value surfaces, gradients can be determined by measuring the spacings between lines of equal value. These measurements provide indices to relative rates of decline or increase in land values as one moves in different directions from different parts of cities. Other distance, area, and directional measurements can be made on maps by employing a variety of symbols. A map of commuting times around U.S. cities, for example, might be superimposed on a population dot map to measure the number of people within different time-distance zones of cities of varying size.

Maps are also useful as devices for the transformation of surface

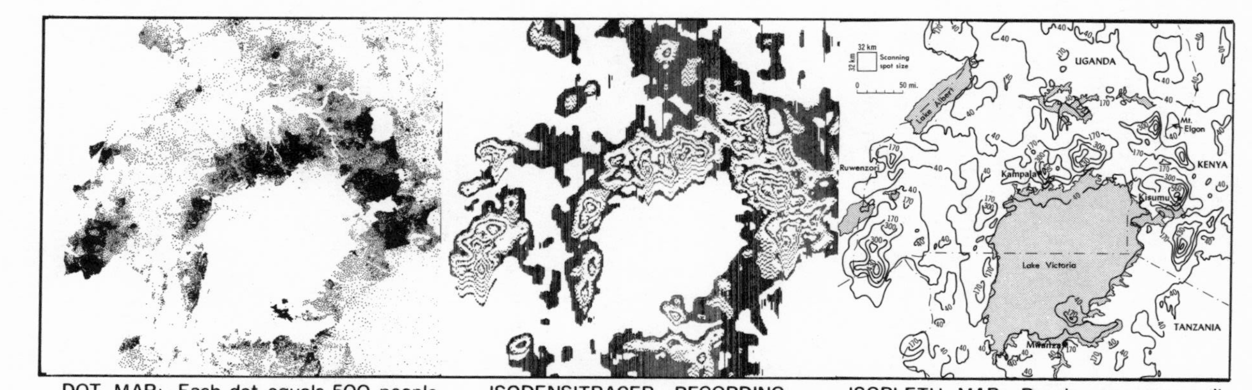

DOT MAP: Each dot equals 500 people ISODENSITRACER RECORDING ISOPLETH MAP: People per square mile

FIGURE 25 POPULATION DOT AND DENSITY MAP

A discrete distribution such as a dot pattern transformed into a continuous isopleth surface by use of the Isodensitracer, an optical densitometer.

Courtesy Philip Porter, University of Minnesota.

characteristics in the plane. Directional and positional relationships, such as linear and nonlinear alignments, great circles, hierarchies of centers, and tributary areas with certain geometric properties, may be derived from maps. Maps serve as information filters by transforming complex patterns into simpler patterns for the purpose of comparing spatial distributions. For example, as shown in Figure 25, one may convert a dot map of population into an areally continuous map of contours enclosing zones of equal population density by using an instrument such as an optical densitometer. Another type of transformation is the substitution of other dimensions shown on maps. Distance may be scaled in a nonarithmetic fashion, as in Figure 26, in which distances from Asby parish in Sweden are scaled

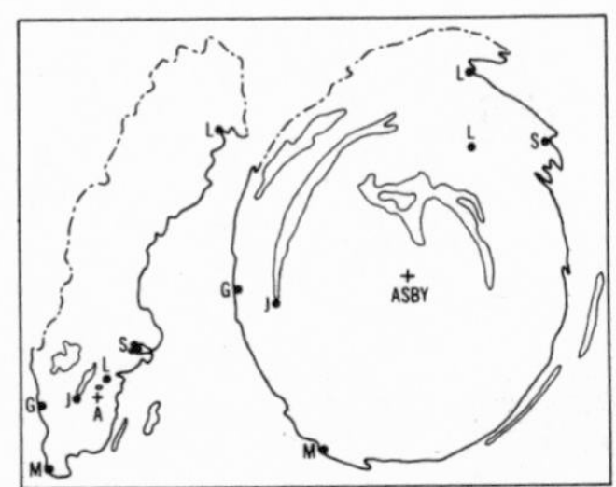

FIGURE 26 LOGARITHMIC TRANFORMATION

The map on the left is a map of Sweden on a standard arithmetic scale. In the map on the right, distances are scaled logarithmically in all directions from Asby. Thus, it is possible to show more detail close to Asby, as in the cases of the lakes. Such a map may be a useful base for plotting data that decline rapidly from a given point, such as phone calls or auto traffic.

After Torsten Hagerstrand, "Migration and Area," *Migration in Sweden*, "Lund Studies in Geography, Series B, No. 13" (Lund: C. W. K. Gleerup, 1957), p. 73.

logarithmically. This provides a more useful base map for measuring migration flows or telephone calls, where the central county records much larger figures than do outlying counties. Area transformations may also increase the utility of a map. Figure 27 is a world map transformed to make country areas proportional to population, thereby providing a base for plotting different demographic measures.

Finally, and perhaps most important, the map serves as a useful device for generating hypotheses and suggesting possible avenues for their revision. Comparison of two maps may lead to a hypothesized relationship as in the case of two maps of land values and popula-

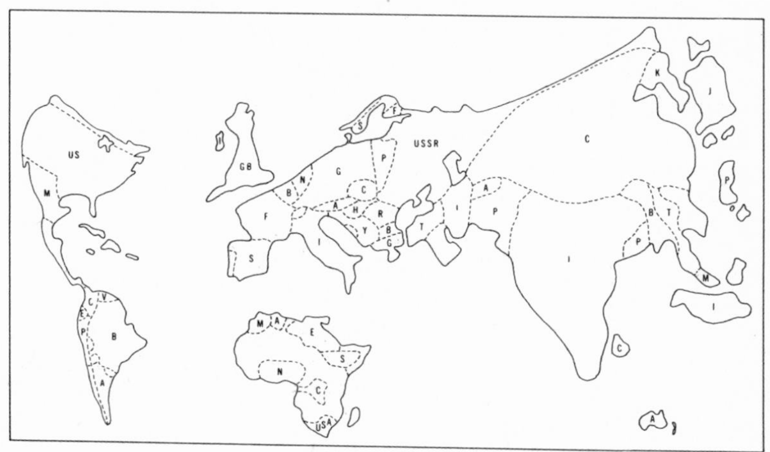

FIGURE 27 POPULATION-BASED WORLD MAP

Areas of each country are proportional to that country's population.

After W. S. Woytinsky and E. S. Woytinsky, *World Population and Production* (New York: Twentieth Century Fund, 1953), pp. 42–43.

tion density. The closeness of the relationship between the two variables may then be tested by curve-fitting or regression procedures and a work map made up displaying departures from these relationships. Examination on this work map of the regression residuals (or the ways in which land values in a particular city are *not* explained by a few relatively obvious factors such as population density) usually suggests other, less obvious, factors which may then be incorporated into expanded and more complex models, including more than two variables. The whole procedure may be repeated until the investigator finds himself unable to associate the patterns of residual variation with other factors. Hypotheses may also be tested by using maps of simulation patterns, as in the examples of diffusion innovation and the process of migration out of the Negro ghetto (Figures 21-A through 21-D). In both these cases, a set of probabilities was derived from decision-rules based on certain postulates describing spatial behavior. These probabilities were used to produce a map on which the distribution of the innovation or of the Negro migrants was simulated according to the probabilities used. This simulation map therefore represented one sample of how the map of Negro population would look if the only nonrandom processes at work were those represented by the postulates chosen. Such maps are then compared with maps of the actual distribution and consistent discrepancies used as the basis for developing new postulates.

The computerization of certain aspects of cartography has already demonstrated great promise for geographic work. Many standard programs for computerized mapping are now available. For example, the Computer Graphics Center at Harvard has developed the popular SYMAP Program (Synagraphic Computer Mapping) and has organized institutes to train geographers and others in its use. A major project is under way in Great Britain to examine the possibilities of the large-scale computerization of certain types of cartographic work. Figure 28-A is an example of a contour map drawn to show the distribution of Roman Catholics in eastern United States. Figure 28-B is a computerized version of a population surface. The simulation studies referred to above are particularly well adapted to map output. Once the model has been fed into the computer, the investigator can quickly obtain a series of maps showing the probable consequences of changes in decision-rules on the land-use pattern. The decision-rules may relate to zoning, new transport facilities, or alternative locations for shopping centers.

The Canada Land Use Inventory is an example of a plan to combine computerized cartography and information systems. Systems of digitizing and locational coding make it possible to store a wide variety of socially relevant data and have it available for quick cartographic print-out or display. It should be possible to obtain rapid print-outs of maps and tabulations that involve different combinations of variables. If, for example, an investigator is looking for a site that meets certain requirements of labor-force availability, proximity to water and certain natural resources, he merely has to feed these requirements into the computer and a map delimiting all areas meeting these criteria will be printed out. Because of recent rapid advances in computer display facilities, particularly cathode-ray tubes, and pattern recognition techniques, the time is fast approaching when the manipulation and analysis of maps may be done on such display facilities.

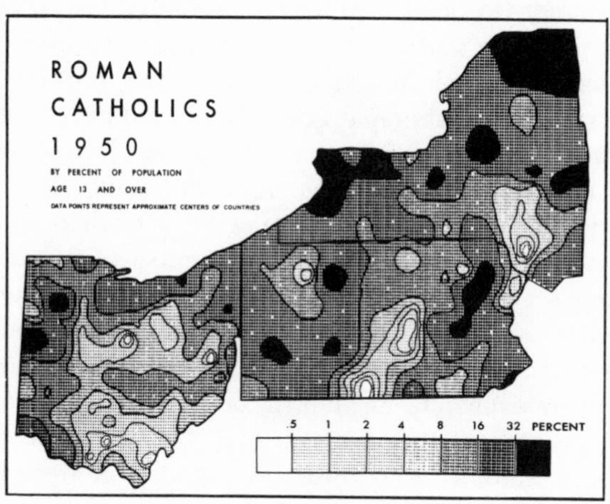

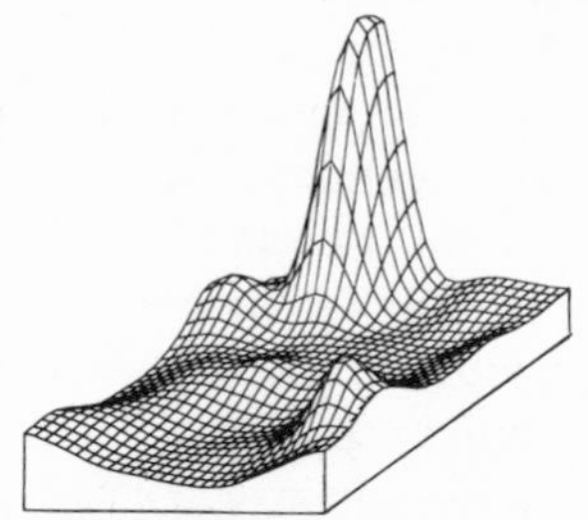

FIGURE 28 COMPUTERIZED MAPS

The map on the top shows the 1950 distribution of Roman Catholics in northeastern United States. The shading within the contour lines was produced by the SYMAP computer program. Courtesy of Stephen W. Tweedie, Syracuse University. Data from National Council of Churches Survey.

The figure below is a three-dimensional drawing representing population density in central Kansas, with Topeka the central value. From *Context*, Newsletter for the Laboratory for Computer Graphics, Harvard Graduate School of Design, Department of City and Regional Planning, February, 1968.

Mathematical and Statistical Techniques

There is a close relationship between cartographic analysis and the mathematical analysis of spatial models. Although it is most common to find both used in a study, it is quite possible to carry on explicitly spatial mathematical research without using maps.

Mathematical work in geography has a long tradition, particularly in cartography, but its rapid growth in economic, urban, and social geography is a relatively recent phenomenon. The United States has been the leader in quantitative work, but some of the original inspiration came from theoretical as well as methodological developments in Sweden and Germany. The number of universities offering quantitative courses in geography increased from three in 1958 to twenty-nine in 1965, and the number of dissertations employing mathematical or statistical analysis increased from one or two in 1958 to approximately one-fourth of the total in 1964 and 1965. The expansion continued through the late 1960s with many younger geographers employing mathematical models in their teaching and research, and the establishment of a new journal to serve as an outlet for mathematical and theoretical work in geography.

The development of mathematical work in geography also illustrates the impact of government and foundation support on a disciplinary trend. Some of the initial work at the University of Washington was given impetus by Social Science Research Council Institutes on Mathematics for Social Scientists. The first symposium on quantitative geography, sponsored by the Geography Branch of the Office of Naval Research in 1960, brought together a group of younger geographers who were at the forefront of current research. Diffusion of quantitative methods into the graduate curriculum was aided greatly by three six-week summer institutes in quantitative analysis for university geographers sponsored by the National Science Foundation in the early 1960s. Advanced Science Seminars in the late 1960s for quantitative geographers, sponsored by the National Science Foundation, concentrated on specialized aspects of the field but continued to have a strong accelerative effect.

The earliest developments consisted primarily of borrowing inferential statistical techniques from other fields, particularly those applicable to area sampling and correlations between patterns observed on maps. Some of the difficulties and particular needs of the analysis of spatial organization became evident as empirical studies progressed, and the work soon evolved toward a variety of statistical and mathematical models applied to many themes common to geographic investigation: the relationship among spatial distributions; classification of areal data into regional systems; the study of processes; the relative efficiency of different spatial organizations; and the study of linkages.

The relationships among spatial distributions were initially treated with regression. Instead of comparing a series of maps, regression equations were developed to express relations between a number of spatially distributed variables. It also became a common procedure, as discussed above, to map and analyze regression residuals in order to suggest further relationships, which might then be incorporated into expanded versions of the original models. Later work dealt with the particularly geographic problem of spatial autocorrelation, the tendency for phenomena located close together to show similarities. Different sorts of procedures, based on ratios to express nearness or contiguity and on cross-spectral analyses to measure two-dimensional correlation, have been developed and adapted to take account of this tendency, but their effectiveness to date has been limited. An interesting variation on the regression theme has been the use of orthogonal polynomials in trend-surface analysis, adapted from physical geography and geology. The locational coordinates are literally incorporated into the terms of the equations, breaking the contour map down into a set of simpler surfaces, as shown in Figure 29. Settlement studies in which space and time coordinates measure changes over an area may also be handled by trend-surface methods. In Figure 30, a complex map of pioneer settlement dates, for example, can be shown to have a high degree of spatial regularity by filtering out a linear then a quadratic surface.

Problems of strong interdependence among spatially distributed variables led to a concern with various forms of factor and principal components analysis. These methods were applied to many problems that involve scales of development and economic health for cities, regions, and countries; urban functions; industrial location factors; characteristics of peasant agriculture; flow patterns of commodities; and the nature of modernization processes within developing countries.

The problem of classification in geography has also been approached with multivariate techniques combined with different

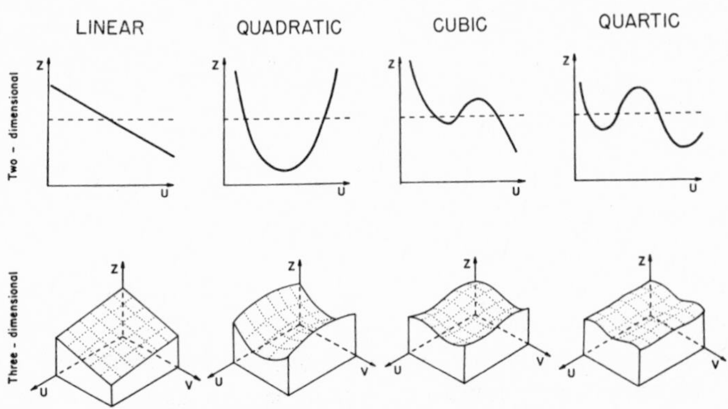

FIGURE 29 TREND-SURFACE DIAGRAMS

Diagram representing relatively simple trend lines and corresponding trend surfaces that may be broken out of complex surfaces as represented on a contour map.

After William C. Krumbein, "Regional and Local Components in Facies Maps," *Bulletin of the American Association of Petroleum Geologists*, XL (1956), 2163–94. From Peter Haggett, *Locational Analysis in Human Geography* (New York: St. Martin's Press, Inc., 1966), Fig. 9.22.

grouping procedures. Different scales were developed through multivariate analysis and used to classify areal units such as countries or census tracts at varying levels of generalization. Grouping procedures that took account of contiguity produced the regionalization schemes. Nonmetric classification methods are also widely available today in programmed form to handle more qualitative information. Some experimental work with animated film has also been carried out on the spatial dynamics of a regional classification. One film shows a progressive grouping of southern counties classified on a large number of socioeconomic variables. The first counties to appear on the map are the most characteristic, or typical, counties, then the progressively less typical counties appear, until those with the least typical southern characteristics fill the last spaces on the map. Film presentation provides an unfolding view of the complex spatial patterns impossible to achieve in the usual static sequences.

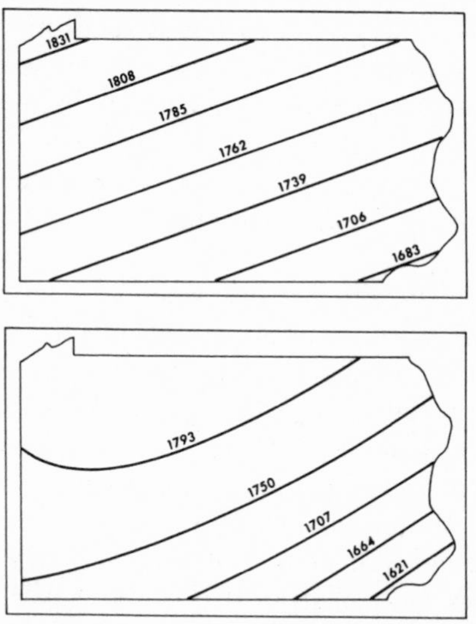

FIGURE 30 SETTLEMENT TREND SURFACES—PENNSYLVANIA

Linear and quadratic surfaces generalized from a complex contour map based on dates of settlement of Pennsylvania communities. Courtesy John Florin, The Pennsylvania State University.

Spatial processes have also been studied with the aid of probability models that include Markov processes, queuing theory, and Monte Carlo simulations. These and other methods have been applied to the development of settlement patterns, port-hinterland relations, migration patterns, urban-travel habits, land-value changes at highway interchanges, patterns of adoption of innovation, and others. Settlement patterns have been studied as sets of points describable in terms of different probability density functions. Attempts are currently being made to interpret the basic equations of these functions according to the spatial processes that bring about different settlement patterns as well as to use these equations to simulate settlement patterns. The complexities of many of the spatial processes as they developed in the changing context of cultural and physical barriers had led to the use of simulation models as discussed above (see pages 28 to 31). One of the more difficult problems currently being dealt with is that of verification. How close does a simulated pattern have to be before it can be considered a reasonably close approximation of the original pattern? Even if the approximation is unquestionably close, there remains the problem of interpreting the meaning of the relationship between the mathematical parameters of the model and the real world processes involved.

An increasing concern with the efficiency of alternative spatial organizations has led to an interest in extremum models of different sorts during the 1960s. The purpose of these models is to maximize or minimize some quantity, such as transportation costs, subject to a set of constraints, such as limitations on productive capacity. Linear programming formulations have been used to study such things as the allocation of flows of a given commodity over a transportation network so as to minimize transport costs and yet satisfy all surpluses and deficits of that commodity; the allocation of units in a set of hinterlands so that each is assigned to the nearest urban node; and the determination of minimum cost diets for different parts of an underdeveloped country. The use of these and similar models represent an interesting departure for the geographer who has traditionally asked why things are where they are, rather than where things *ought* to be in a normative sense. In some instances, however, the "why" question is still asked, and the optimal solution is used only as a partial explanation of the observed pattern. Discrepancies between observed patterns and optimal patterns are then used as the basis for further investigation. A more formal reaction to the relatively restricted and highly rational value system implicit in the earlier linear-programming formulations has been the more elaborate formulation of constraints, and the experimentation with variations such as those, that incorporate changes through time.

The study of spatial interaction has been given impetus by the ease with which transport, trade, and communications linkages may be expressed in the language of graph theory. Studies have included: the use of aggregate graph theory measures to compare national transport networks; the evaluation of accessibility of individual nodes (cities) to an entire network; the variation of accessibility through time and with mode of transportation; the evaluation of trade linkages between countries; and the nature of the urban hierarchy as defined by different networks. Recent developments in methodology permit the more critical analysis of flow systems and the ability to estimate optimal network configuration for maximizing flow capacity.

Geographers are also carrying on experiments in the application to spatial problems of game theory, spectral analysis, analogue simulation, topology, and certain non-Euclidean geometries. These and other analytical developments, together with the computerization of cartography, will accelerate the research in these fields during the next ten years and will have an increasingly significant impact on the critical analysis of spatial problems at the city, regional, national, and international levels.

Field Methods and Remote Sensing

Geography in the United States as well as in other countries has long been concerned with field study whereby investigators go directly to the original source of data and confront the phenomena to be studied as localized associations in all their spatial complexity. Methods for the field study of small areas have been stressed in U.S. geography since a series of conferences held in the Midwest between 1910 and 1920. At one time it was felt that there was a standard set of geographic field techniques applicable to most problems, but today most geographers agree that the field techniques used in any research project must be those most appropriate to the questions being asked. The nature of these investigative methods is constantly shifting as research emphases shift. The use of the plane-table and fractional-code field maps were once considered to be indispensable parts of any field investigation. Today, there is greater stress on interviewing techniques, areal sampling procedures, problem orientation, and the use of some forms of remote sensing information, which will be discussed further. Thus, the problems of field work in geography have come to resemble those in the other partially field-oriented social sciences such as anthropology and sociology. Much of the work in sectors of geography undergoing rapid development, such as urban and settlement studies, requires rigorous formulation of questionnaire schedules, training in interviewing, and experience in behaving as participant-observer in situations in which external factors can inordinately influence the types of information gathered and the ways in which they can be used. These problems will become more acute as geographic research becomes increasingly involved with behavioral issues like those in the environmental perception field. Field work also plays an important role in geographic education, where it provides students with valuable, first-hand experience in relating the collection and analysis of data to the formulation of both theoretical and real-world problems. The Commission on College Geography of the Association of American Geographers recently issued a report on field training in geography as the first in a series of technical papers.

Recent technological development in remote sensing constitutes an enormous potential advance in the gathering and mapping of field data. Remote sensing is the acquisition of information about objects or phenomena that are not in contact with the data-gathering device. The use of aerial photography, one of the earliest forms of remote sensing, has been widespread in geography since World War II, and it has been particularly important in studies of rural land use and settlement distributions and patterns. However, aerial-photographic-interpretation techniques have never been developed adequately for research on urban complexes, and much further technical and methodological development is needed. Use of aerial photographs is severely restricted, moreover, by security restraints on their use for areas other than the United States.

Recently, there has been a remarkable expansion of technological capability because of the parallel successful development of new types of advanced remote sensors, better aircraft and satellites, and more elaborate systems for transmission, storage, and display of data. Infrared sensors have been developed to pick up variation in heat transmission. These sensors can produce images in which diseased trees appear in sharply different colors, even though no differences would be apparent on ordinary color film. Radar and other scanning devices can produce detailed images regardless of weather or time of day. Multiband analysis permits the comparison of a set of simultaneous color photos of the same place emphasizing different portions of the electromagnetic spectrum. From these, keys can be devised for quick identification of crop types, house types, and certain traffic indicators. The use of satellites permits quick viewing of large areas at frequent—even daily—intervals, and improved resolution makes it possible to enlarge details to such a point that much mapping and observation are feasible at large and medium as well as small scales. Coupling remote-sensing technology with other devices such as television relays, computer storage and retrieval, graphic display, and image-processing devices allows a vast expansion of localized data that records both spatial patterns and processes.

To evaluate the possible geographic uses of this new technology, a conference sponsored by NASA and the Office of Naval Research was held in Houston in 1965, and a long list of topics was suggested in the published report of the conference. It was noted that inventories of a wide variety of phenomena, such as air and water pollution and areas of potential crop failure or urban blight, would be feasible as would internationally comparable resource and land-use surveys at uniform scales. Surrogate measures or secondary indicators were suggested for a number of socially significant phenomena. House types and density and their relation to age, for example, can be matched to population and income characteristics in certain test areas and times, and then used to derive estimates for other areas and times. Growth and change measures can be derived from repeated observations and can be applied to urban expansion, the diffusion of innovation, the spread of plant disease, water pollution, flood hazard, and the monitoring and control of traffic.

Recent work by geographers indicates that population census-taking by remote sensing may be feasible. Certain relationships between city size and area make it possible to estimate population distributions in underdeveloped areas with a fair degree of accuracy.

Interest has been mounting in the United States, and, in addition to the Houston conference, there has been a series of symposia and institutes on remote-sensing. A group of committees on remote sensing, including one for geography, has been established within the National Academy of Sciences. A Geographic Applications Program was established in the U.S. Geological Survey in 1966 and a series of geographic studies are now under way. Some of these studies investigate "ground truth," or the validation of sensor techniques by comparing results against known ground data, and some investigate the adaption of new analytical methods to remote-sensing data. The A.A.G. Commission on Remote Sensing, sponsored by the Chief Geographer's Office of the U.S. Geological Survey in cooperation with NASA, has proposed an experimental plan whereby the TVA area would be covered with a full array of remote sensors and the results compared with the detailed ground material previously available, so as to clarify the processes of change and development of regional-resource patterns.

There are several barriers to the full utilization of present technological capability. The equipment is costly, and the utility of the pictures must be clear before investment can be justified. Tentative plans were made to launch an earth-resources satellite in 1969, but now it is doubtful that they will be carried out. Security is another problem. Many of the instruments have been developed by the Department of Defense, and the restrictions on the use of pictures are, in some instances, both severe and inconsistent. The international implications of data returned from earth-orbiting satellites also pose a difficult problem. An interesting example of international cooperation in this respect, however, is the weather satellites, TIROS and NIMBUS, that have been collecting weather data from orbit since 1960. The World Meteorological Organization provides a framework for setting standards and arranging exchanges. The data are unclassified, the United States and the U.S.S.R. are exchanging information, and the Weather Bureau is sending daily cloud-cover photos to many countries.

SELECTED RESEARCH DIRECTIONS

The four research directions discussed below—locational analysis, cultural geography, urban studies, and environmental and spatial behavior—represent the growing research interests in contemporary geography. All the themes treated briefly in the illustrative studies section of Chapter 1 are included in at least one of the subfields discussed here. Two of these four subfields—locational analysis and cultural geography—represent major research clusters within the discipline as a whole. The other two may be regarded, in part at least, as derivative from these. Thus, the urban-studies interests of geographers perhaps draw most heavily upon the theory and expertise developed within the locational-analysis research cluster, although they also draw from major conceptual resources in cultural geography. Similarly the studies in environmental and spatial behavior reflect important developments in research on perception and landscape in cultural geography, but also draw on techniques and models developed in locational analysis research as well as upon work in other subfields (such as political and physical geography), and in related disciplines such as psychology. Moreover, the locational analysis and cultural geographical clusters are themselves by no means readily separable.

The various physical-geography fields such as geomorphology and climatology are excluded by our terms of reference. Space limitations prohibit meaningful treatment of other research clusters such as political, social, regional, and historical geography. However these studies do interweave with the four research directions discussed, particularly historical and regional geography. Most of the topics treated could be, and in many instances are, viewed from a time perspective, as in the case of nearly all process or sequential studies. Regional geography is equally pervasive, since virtually all the research themes illustrated may be meaningfully applied to the study of specific regions. The concepts and models discussed in the locational-analysis and urban-studies sections may be applied to other countries, and this holds particular research promise, both for generalizing and improving the models and for providing more insight into the spatial organization of different parts of the world.

Locational Analysis

The fields of geography referred to as economic, transportation, and urban geography underwent profound change in the United States during the 1950s and 1960s. In the 1965 report, *The Science of Geography*, the term "location theory cluster" was used to identify the workers in this field. Later, the term "locational analysis" was also used. Since 1965, the work in locational analysis has expanded widely in geography as well as in regional economics and regional science, and planning. In the late 1960s, a series of books summarizing portions of this work was published, and courses dealing with locational analysis began to appear in geographic curricula. This transformation represented an important break with the traditional work in these fields, although it was evident in the literature as far back as the 1930s.

Traditional economic geography was largely production-oriented. The subdivisions were organized around manufacturing and agriculture, and numerous textbooks and articles described the distribution of various types of production. Relations with economics and the use of economic theory were minimal, and the spatial distribution of production was interpreted within a simple framework of natural conditions: climate and soils, in the case of agriculture; location of major resources, in the case of manufacturing.

More critical inquiry into the reasons for the location of various types of economic activity, production, consumption, and marketing resulted in closer relations with economists and others. Geographers began to approach the study of spatial interrelationships more analytically, examine the historical development of particular economic distributions, and relate them to a wider range of factors—social, cultural, and political, as well as physical. The work of the classical location theorists, both economists and geographers, was reexamined and it formed the basis for empirical studies. The work of Walter Isard and the Regional Science Association was important to this reexamination and to the formation of new theoretical bases for locational study. In the late 1950s the use of normative models such as the various linear programming formulations in locational study received greater attention. By the late 1960s studies had started to branch in an interesting variety of directions: individual spatial behavior, explicitly geometric models, probabilistic models, network analyses, general systems analyses, relation to political decision-making, and combinations of these with evolving forms of the original normative models.

Locational Explanation. Map studies of spatial interrelationships were essential to the first critical inquiry into explanations for the location of economic activities. Comparisons of maps depicting the evolution of agricultural activities with a wide variety of cultural and economic patterns exposed the inadequacy of former, physically biased explanations of location. Similarly, in the study of manufacturing, locational patterns and their relationships to a wide range of factors were examined. A number of statistical studies were made of the areal association between these factors and the location of different activities.

Beside the increased concern with locational explanation was a more explicit concern for the ideas of such classical location theorists as von Thünen, Weber, Lösch, and Christaller. In agricultural study more specific attention was given to the way in which land values and the intensity of farming decline as distance from the city increases. Empirical studies showed complex relations between these relatively simple distance-based models and complicating factors such as agglomeration, cumulative causation, and cultural barriers.

Manufacturing studies were more specifically treated in the work of Alfred Weber, which dealt with the location question as viewed by an individual firm and stressed the minimization of transport costs for differing combinations of raw material and products. Figure 31, for example, shows that Monterrey is the point of minimum transport cost for iron and steel production in Mexico. The contour lines (isodapanes) show how much more costly it would be per ton of finished steel to locate anywhere but the minimum point. At Mexico City, for example, it would cost 30 pesos per ton more, so that such a location might be optimal only if some other cost, such as labor, resulted in a savings of more than 30 pesos per ton as compared to Monterrey. Figure 32, shows the zone of minimum transport costs to market in the United States and is derived from a study of the single factor of overall proximity to a national market based on total retail sales.

Work in fields such as transportation and the study of commercial activity also proceeded toward more general considerations, which revealed their affinity to the other work in locational analysis. The work of Edward Ullman and others on spatial interaction as evidenced in port-hinterland relationships, railroad networks, and commodity-flow studies led to the use of gravity models to predict the flow of traffic between points as modified by the effects of intervening opportunity and complementarity of production. The classical work of Christaller on central-place theory, that sets forth a schema for a hierarchical pattern of settlements and market areas (Figure 33), was subjected to a series of empirical tests and modifications as work dealing with commercial activity proceeded. (See the discussion in the section, "Urban Studies," below.)

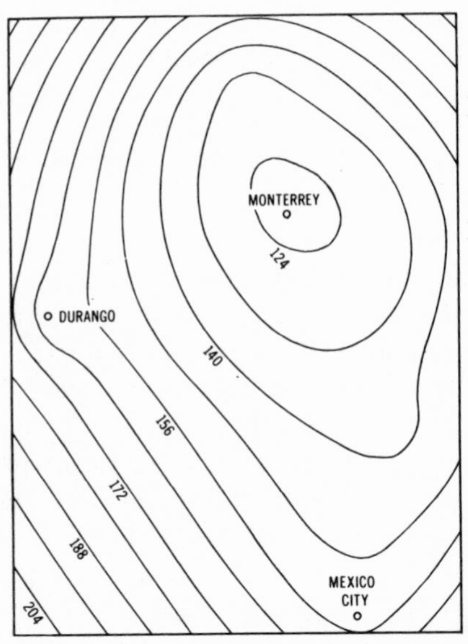

FIGURE 31 ISODAPANES: MEXICAN IRON AND STEEL

Isodapanes, or contour lines based on weight, distance, and freight rates, showing how much more costly it would be per ton of finished steel to locate at points other than the point of minimum transport cost (Monterrey).

After Robert Andrew Kennelley, "The Location of the Mexican Steel Industry," *Revista Geografica,* XVII, No. 43 (1955), 60–77, Fig. 11.

In the late 1950s and early 1960s the publication of a series of expository articles by William Garrison and several books by regional scientist Walter Isard called attention to the growing body of literature that deals with the application of various forms of linear programming and other optimizing models to locational problems.

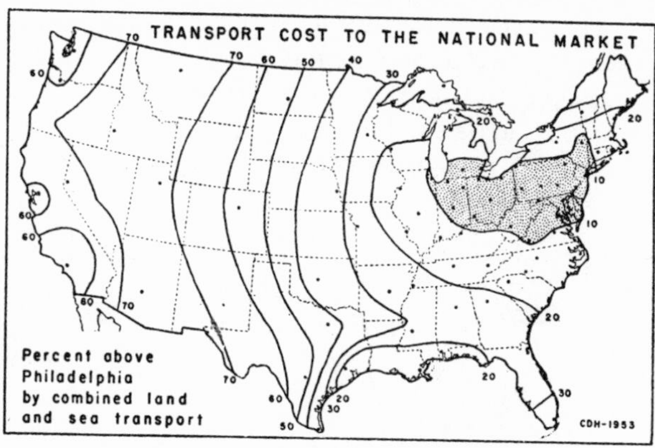

FIGURE 32 TRANSPORT COST: U.S. MARKET

Contour lines show variations in transport costs for reaching the entire U.S. market from major cities. Philadelphia has the lowest costs for reaching the entire U.S. market assuming that shipments to each county are proportional to the retail sales of the county. Contours indicate the per cent above Philadelphia for each zone.

From Chauncy D. Harris, "The Market as a Factor in the Localization of Industry in the United States," *Annals*, Association of American Geographers, XLIV, No. 4 (1954), 315–48, Fig. 9.

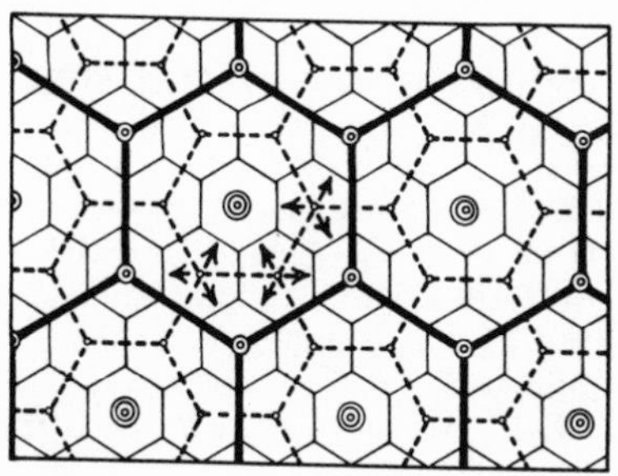

FIGURE 33 THEORETICAL SETTLEMENT PATTERNS

Idealized settlement pattern shows a hierarchy of centers and tributary areas with lower-order centers and areas nested within higher-order areas.

After Walter Christaller from Barry J. Garner, "Models of Urban Geography and Settlement Location," Chapter 9 in *Models in Geography*, ed. Richard J. Chorley and Peter Haggett (London: Methuen & Co. Ltd., 1967), pp. 303–60.

Since then, geographers, together with economists, regional scientists, and members of related disciplines, have been dealing more explicitly with questions of the optimality or efficiency of differing spatial organizations. Some of the classical locational work may be readily fitted into such formats. For example, if one is given a set of raw-

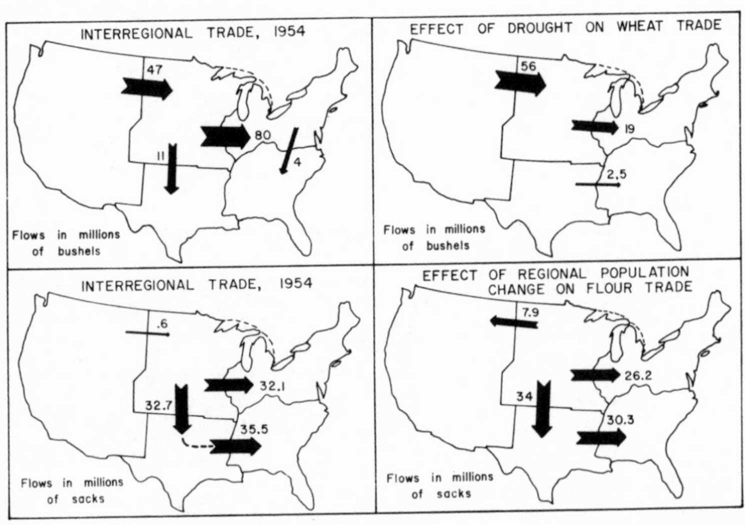

FIGURE 34 OPTIMAL FLOW PATTERNS: WHEAT AND FLOUR

The two maps on the left show interregional trade patterns for wheat and flour under conditions of minimum transport cost. Those on the right show the effects of drought and population shifts on optimal trade patterns for wheat and flour, respectively.

After Richard L. Morrill and William L. Garrison, "Projections of Interregional Patterns of Trade in Wheat and Flour," *Economic Geography*, XXXVI, No. 2 (April, 1960), 116–26, Figs. 1, 3, and 4.

material sources, factories, and markets, he may determine the flow pattern that will meet market demands at a minimum cost without violating the capacity limitations on factories and raw materials. Figure 34 shows optimal flow patterns for wheat and flour under existing conditions, as well as under certain assumptions for drought and regional population shifts. Figure 35 shows the location rent surface for wheat farms in the upper Midwest based on the value accruing solely from location relative to international and domestic wheat trade. In other studies optimal boundaries were developed for the distribution of both medical services and administrative services, and their asociated tributary areas.

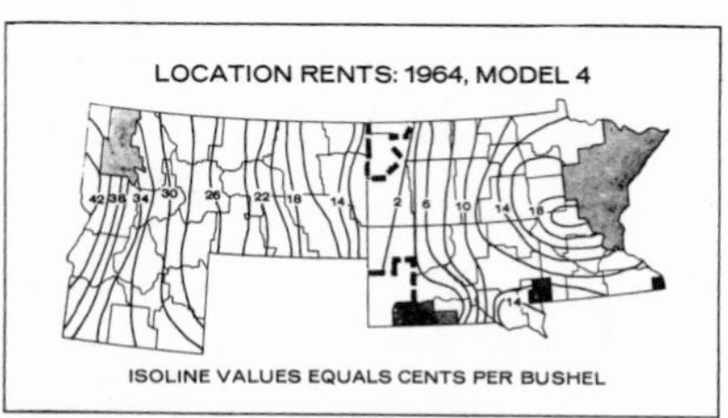

FIGURE 35 LOCATION RENT

Location rent values for wheat farms in the upper Midwest. Location rents represent values associated solely with location relative to international and domestic wheat trade.

From Donald W. Maxfield, "Spatial Analysis of Location Rent," paper presented at annual meeting of Association of American Geographers, Washington, D.C., September, 1969, Fig. 4D.

Behavioral Emphases. By the late 1960s, locational analysis had expanded in many directions, some closely related to previous work, some representing new experimental thrusts. Some of the normative models were brought into a more explicitly behavioral context. A study of farming in Sweden, for example, used a programming model to establish the maximum income surfaces that could be obtained if totally rational decisions were made on the basis of complete information. The problem was then to examine why actual distribution of agricultural production differed from the optimal pattern (Figure 36). Further investigation revealed differing attitudes toward risk, and varying spatial intensity of information flows about new farming techniques. (See the discussion of diffusion of innovation below, pp. 69–70. Figure 37 shows a behavioral matrix developed by Allan Pred for application to locational models. The ability to use information effectively increases across the row; knowledge increases down the column. Individuals located in the lower right-hand corner of this matrix would tend to locate closest to the optimum position predicted by the model since both their knowledge level and their ability to use knowledge effectively are at a maximum. At points in the matrix that are increasingly distant from the lower right-hand corner there is an increasing tendency for individuals to locate in positions farther from the optimum, either because of the lack of knowledge or because of inability to use their knowledge effectively.

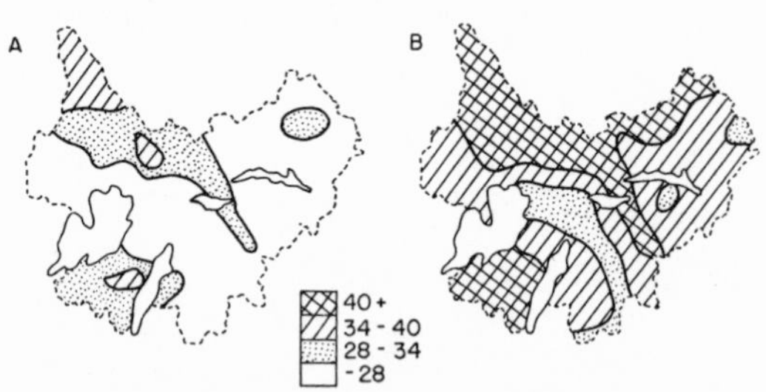

FIGURE 36 DEPARTURE OF ACTUAL FROM OPTIMAL AGRICULTURAL PRODUCTION PATTERNS

Map of actual and potential farm labor productivity. Figure 36-A represents actual farm labor productivity; Figure 36-B represents the optimal productivity if all resources were allocated in the most efficient fashion.

After Julian Wolpert, "The Decision Process in a Spatial Context," *Annals*, Association of American Geographers, LIV, No. 4 (1964), 537–58, Figs. 2 and 3. From Peter Haggett, *Locational Analysis in Human Geography* (New York: St. Martin's Press, Inc., 1966), Fig. 6.18.

Other programming studies that introduce stochastic and other complex constraints and that experiment with variations such as recursive programming to give more recognition to the evolutionary nature of most economic distributions are being carried on by geographers and economists. Work is also being carried on with models based on individual behavior in space, as in studies of search patterns and the application of learning theory to consumer behavior.

Spatial Emphases. The appearance in the late 1960s of *Locational*

Analysis in Human Geography, by the British geographer Peter Haggett, and *Models in Geography*, by Haggett and Richard Chorley, was symptomatic of a more consciously spatial emphasis in locational analysis. A number of U.S. and British spatial studies were sum-

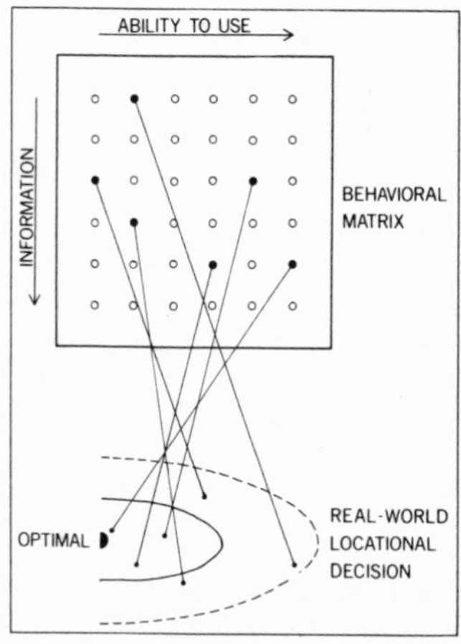

FIGURE 37 BEHAVIORAL MATRIX

Diagrammatic representation of the relation between optimal location and selected behavioral characteristics. For individuals located near the lower right-hand corner of the matrix, locations are closest to the optimal, since they have much information and a high ability to use it. Individuals near the upper left-hand corner of the matrix should locate farthest from the optimal, since they are near the lowest points on scales of both information and ability to use it.

After Allan Pred, *Behavior and Location*, "Lund Studies in Geography, Series B, Human Geography, No. 27" (Lund: C. W. K. Gleerup, 1967), Fig. 5.

marized, in both published and unpublished papers, and several organizational structures for the field were suggested. Two of the more explicitly spatial themes were network analysis and stochastic process analysis, involving probability models.

Network analysis grew out of the earlier transportation and spatial interaction studies. Graph-theoretic methods were used to study the accessibility of individual points in a network, the overall effectiveness of different network configurations, and the hierarchy of urban centers. In a study of the growth of the Brazilian highway network for example, the matrix of connections and accessibility of each city in the network was computed for each year. Changes in accessibility were compared to urban growth during the same period, and it was possible to make some estimates of the effects of improved transportation (lowered transport costs) on urban growth.

The work of Michael Dacey in the analysis of point patterns and their relation to stochastic processes represents another strongly spatial field of experimental study. Some insight has been gained into the nature of settlement processes and the location of cities by noting the close conformity through time of actual patterns to patterns predicted by certain types of probability distributions. For example, a number of opposing forces might produce a random pattern; a contagious process might produce a pattern more clustered than random; a competitive process might produce a pattern more regular than random. Models have been developed to fit complex spatial series, and efforts are being made to interpret the mathematical processes generating these models in terms of real-world spatial processes.

Systems Emphases. The work in locational analysis has also broadened in scope. Brian Berry and others carried the study of commercial activity well beyond the postulates developed by Christaller and attempted to synthesize central-place regularities and other regularities related to such things as the rank and size of cities into a general systems framework. Political considerations have also assumed a more significant role in locational analysis. Research concerns include the consequences of differing systems of administrative areas, the effects of political boundaries or differing ideologies on spatial organization, and the complex interplay between economic and social optimality, on the one hand, and political feasibility, on the other, in the formation of public policy. As locational concerns progressed from a relatively narrow economic base toward more social, cultural, and political considerations, the relation of locational to regional analysis became more evident. For example, one study attempted to simulate the development of a region (shown in Figure 38) where transportation and urban and industrial develop-

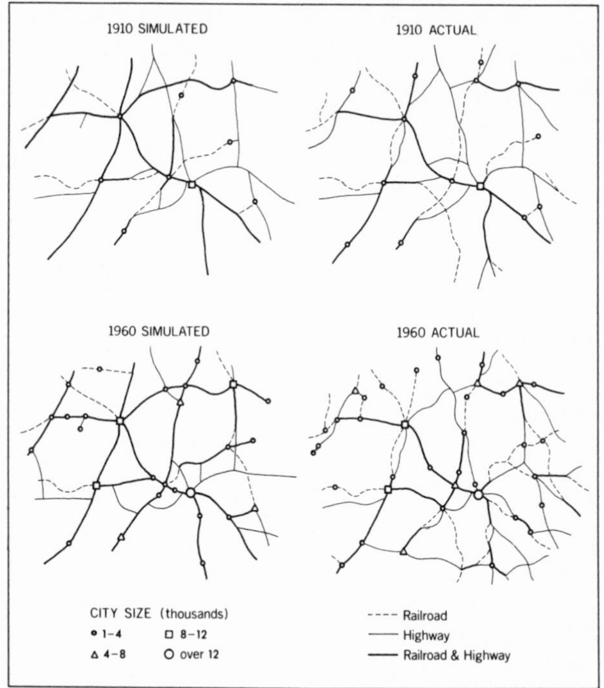

FIGURE 38 REGIONAL SIMULATION

Simulated and actual patterns of regional development. The growth of urban centers, industries, and transportation links was simulated sequentially with attempts made, at each stage, to include the effects of these phenomena upon one another.

After Richard L. Morrill, "The Development of Spatial Distributions of Towns in Sweden: An Historical Predictive Approach," *Annals*, Association of American Geographers, LIII, No. 1 (1963), 1–14, Figs. 2, 5, 7, and 8.

ment were simulated sequentially, and some attempt was made to encompass the feedback effects they had on each other. A more formal and comprehensive model of a similar sort, developed by Leslie Curry, involves a dynamic central-place system, a locational decision mechanism, and diffusion through interaction in a single schema that relates behavioral processes to spatial structure.

The importance of general systems approaches, including feedback, was stressed in *The Science of Geography*, a report of the National Research Council ad hoc Committee on Geography. English geographers have postulated open systems approaches to describe equilibrium conditions of the physical landscape, while others have postulated systems analysis as a unifying methodology in human geography. Researchers have noted the relationship of geographical systems to ecosystems, and there have been some studies that treat the interlocking biological, physical, cultural, and physiological systems of certain primitive groups. The Commission on College Geography has recently sponsored a course outline by two Canadian geographers, that presents a general systems approach to an undergraduate economic geography course.

There has by no means been a steady progression of work in locational analysis. Most of the themes mentioned above are still being pursued in a variety of forms, and there is much that is clearly experimental about the more recent investigative models. There have also been parallel but promising developments in mathematical regionalization and the application of locational analysis to historical geography.

It is probable that the diffusion of theoretical knowledge—as in locational analysis—has been slower among academic geographers than has been the diffusion of quantitative techniques, due in part, perhaps, to the persistent confusion between the two. In the universities there is still a disparity between the increased use of theory at the research level and the meager use of theory in the undergraduate courses, the teaching of which occupies so much of the geographer's time. Relations with other behavioral and social scientists and with analytically inclined planners are improved but are still slow in developing. Many geographers are still deficient in their training in basic theory in allied fields, and many behavioral and social scientists are still unaware of contemporary geographic work.

Cultural Geography

Cultural geography as the study of relationships between landscape and culture has been distinguished by its interest in continuities from the past, its rejection of environmentalist explanations of social phenomena, and its emphasis on cultural pattern as the key to process. From its comparative vantage point, cultural geography attempts to determine what identifiable forms of social organization, what technologies, what systems of belief, have played their roles in the evolution of landscapes. Traditionally the emphasis has been on historical cases, and the recovery of evidence on these has constituted a chief interest of cultural geographers.

The spectrum of cultural geographic study runs from work on concrete communities in all their aspects, through work that concentrates on particular social institutions, processes, and products, to investigation of given technical-ecological arrangements like farming. It is further enriched by a cross-cultural application of new theoretical constructs. It is characterized by a comparative viewpoint that has worldwide scope and unrestricted historical depth. An equally important (if still largely implicit and unspoken) diagnostic feature of cultural geography is its aspiration toward a comprehensive (universal and consistent) vision of man on earth that will embrace all societies and all ecologies.

The Roots in Intellectual History. The attitudes that men have held at different times about the world around them have been a topic for important scholarly investigations by a number of cultural geographers. Among the most distinguished contributions of recent years, *Traces on the Rhodian Shore* by Clarence Glacken, examines the developing conceptions of environment from the Greek philosophers to the major thinkers of eighteenth-century Europe. The first of two expected volumes, the work records the influence of nature on human character and custom, the vision of the earth as a perfect home for man, and the deeply held Western aspiration to achieve control of nature by rational means. In the same tradition, though of more restricted scope, is the recent study *The Hydrologic Cycle and the Wisdom of God* by Tuan. This book deals with the efforts of Renaissance and later thinkers to relate their ideas of an orderly universe to their knowledge of the circulation of the atmosphere, the movement of the seas, and the properties of the climates.

These contemporary works are but continuing steps in a long sequence that summarizes and interprets man's knowledge of the earth at various historical periods. They continue the theme of intellectual history that was earlier expressed in *Geography in the Middle Ages* by Kimble, and *Geographical Lore at the Time of the Crusade* by Wright. These books have contributed strong scholarship to an area that is still vital for future research.

Cultural Ecology. The concern with cultural patterns and cultural processes embraces both ecology and spatial diffusion. While a large number of monographs and articles have been published on particular problems that involve cultural patterns in various parts of the world, only a few integrative and synthesizing studies have been attempted at the world scale. Notable among these is the comparative survey of *Shifting Cultivation in Asia* by Spencer, together with his continuing research on the origin of agricultural regions. Similarly, studies on the distribution of the habit of milking animals in the Old World have indicated that substantial areas of Africa also lack such a trait. A further study surveyed the surprising variety of food prejudices, since these prejudices could affect the rational improvement of nutritional standards in poor countries. A number of geographers have worked on problems of nutrition and food supply that involve cultural differences. This is particularly true for Central and South America.

Cultural geographers have also paid close attention to the distribution of religious attitudes. In the United States Meinig has delimited the Mormon culture region, and has indicated the frequent importance of religious groups and attitudes in distinguishing the subcultural units of Texas. The importance of religious activities to the total cultural system of an area has recently been summarized in *The Geography of Religions* by Sopher. This general work draws heavily upon the author's research in India, published in a sequence of articles over the past few years. Particularly noteworthy is the analysis of the spatial patterns of pilgrimages that demonstrates the way in which frequency of religious movements vary with illiteracy, the relative attraction of sacred and secular places, and the anomalies between actual and generated distributions of religious movements within a circulation system. Sopher's studies reveal some interesting correspondences between pilgrim circulation fields and cultural subregions, as well as significant regional variations in level of commitment to particular religious traditions. Figure 39, a map of residuals from regression (see p. 45), that shows departures from the numbers of Jain pilgrimages to a shrine area that would have been expected from various areas according to their respective distances from the shrine and their respective Jain populations. The resulting pattern of positive and negative anomalies shows a weaker com-

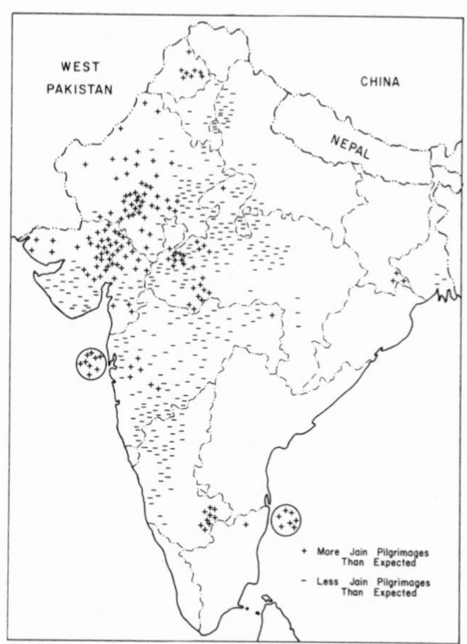

FIGURE 39 RELIGIOUS PILGRIMAGES: INDIA

Departures from the numbers of Jain pilgrimages to a shrine area that would have been expected from various areas according to their respective distances from the shrine and to their respective Jain populations. Areas with minus signs are those in which there is a weaker commitment of certain subgroups of the Jain population to an orthodox branch of their religion.

From David Sopher, "Pilgrim Circulation in Gujarat," *Geographical Review,* LVIII, No. 3 (July, 1968), 392–425, Fig. 11.

mitment of certain subgroups of the Jain population to an orthodox branch of their religion, and thereby provides a possible index to a cultural regionalization based on religious behavior. These studies are a part of the larger problem of understanding sociocultural fields and the consequences of the discordances between these fields and the fields defined by economic transactions.

In their descriptions of the cultural dynamics of regions, geographers have frequently focused upon land settlement. Following the important collection of papers by Bowman, *The Pioneer Fringe,* and the studies by Pelzer, *Pioneer Settlement in the Asiatic Tropics,* the record of settlement as an expression of culture has been examined in many parts of the world. The problems treated range from careful discussions of resettled agricultural colonies in Latin America to work on urban migration streams, and are taken as a set of responses to various attractions and repulsions in the migration fields of urban nodes. Clark has developed the historical geography of the maritime provinces of Canada, and indicates the way in which the people of the maritime provinces stubbornly express the variety of their heritage in the present landscape.

Ecological studies reveal the complexity of the way in which man's cultural heritage shapes, and is shaped by, the total ecology of an area. The idea of ecosystems has given impetus to such study, and both geographers and anthropologists who deal with cultural ecology have shown a convergence of themes during recent years. A major symposium, *Man's Role in Changing the Face of the Earth,* was devoted to an ecological theme, and numerous articles on man's integration with his environment have appeared in recent years. Geographers have been able to reveal the insidious natural and human processes of soil degradation well before they had become serious problems. For example, a study of northern Morocco demonstrated the catastrophic effects of timber cutting and sheep grazing on the forests and grasslands of that country. An ecological-systems approach, however, is not confined merely to the analysis of the deterioration of the physical environment. This approach has also been used by cultural geographers when focusing upon a whole human community. A cross-cultural study, for example, although based on limited evidence, suggests that the urbanization process in Iran differs but little from that in western Europe and North America. The cities and villages of Iran serve much the same functions as do those in the other two areas, and display similar relationships to their surrounding hinterlands. An ecological study of a Chinese village in Malaya noted the barriers to economic development posed by the contrasting views of the Chinese peasant and the Malayan government as to what constituted reasonable norms for agricultural practice and their differing perceptions of the regions within which their activities should be administered. Wheatley's study of the origin of terracing, based on literature from all the social sciences as well as history and sinology, may prove to be a major breakthrough in understanding the relations among agriculture, nature, and culture in Asia. Sometimes the results of such research are of interest primarily for historical reasons, like the ecologically revealing studies of the coastal area of Baja, California, or the studies of pioneer colonization in Antioquia. While the latter described the historical formation of a Columbian highland community of a distinctive peasant type, it also illuminated contemporary processes of the same nature in other, culturally allied areas. For example, a knowledge of these historical processes has been valuable for a study of the highway as an agent of change in a Mexican *municipio*, where land values, crop patterns, and ethnic relations were influenced by a new road through the area.

Diffusion of Ideas. Diffusion as a source of new ideas has traditionally provided one explanation for the development of cultures—the alternative being independent invention. Many cultural geographers have contributed to the knowledge of origins and interactions of cultures of the past; a foremost classic among such studies is Sauer's *Agricultural Origins and Dispersals*, which interpreted a vast body of literature and outlined a history of farming as derived from a few major centers of origin in both the Old World and the New World. In this work Sauer laid down a basic division between crops planted from seed and those grown from vegetative shoots or cuttings—differences that are still important to the distribution and historical positions of farming systems.

The work on cultural diffusion by Kniffen and Stanislawski has been prominent in the United States. Based upon thirty years of field work, the research by Kniffen on house types as indicators of cultural movements has demonstrated the way in which the present landscape of Louisiana was derived from midwestern, French Caribbean, and southern sources. Stanislawski in "The Origin and Spread of the Grid Town" traces the expansion of the square street grid, so familiar to the American scene, from the ancient Near East and

Indian sources via the Romans and Spaniards to the New World.

In Meinig's study, *The Great Columbia Plains,* the diffusion of settlement into an area is treated as a spatial process related to the Turner frontier hypothesis. Figure 40-A shows the first stage, in which fur traders occupied a set of selected points already occupied by missions and army posts; Figure 40-B shows how ranching then came in to spread areally outward, primarily from a zone around Walla Walla; Figure 40-C shows the strongly linear impact of the railroads, which served to establish a second dispersal center at Spokane Falls; and Figure 40-D shows another areal expansion as dryland agriculture moved out from a number of points (including those established by the railroads).

Cultural diffusion has recently received considerable attention from scholars, who are using mathematical methods for the first time. The research of Hagerstrand, Morrill, Brown, Pitts, and others on migration and the spread of innovation is linked strongly to the work in locational analysis and urban studies. It is clear that the ideas and practices that govern a people's relation to their environment change over time, and that an important element in such change is the exchange of knowledge and opinions. The slow and steady spread of ideas, or artifacts embodying them, has been invoked to account for the expansion of rumors and fashions and, more significantly, for the spread of the elements of modernization. The impact of the United States and other Western countries on the so-called "underdeveloped world" is due, in part, to this kind of phenomenon. Nevertheless, the diffusion process as a spatially ordered series of events has not been fully understood in all its possible forms to date. To model the complex processes underlying the spatial patterns of diffusion, a number of younger scholars have turned to computers and other modern aids to overcome the difficult mathematical problems encountered in diffusion processes over heterogeneous areas.

Cultural Perception and the Aesthetics of the Visual Landscape. Cultural geography has been preoccupied for many decades with understanding how ecological relationships in non-Western societies are seen by members of those societies. Its interpretation of cultures and landscapes has rested on a certain openness of mind and a skeptical disposition toward the familiar, and supposedly rational,

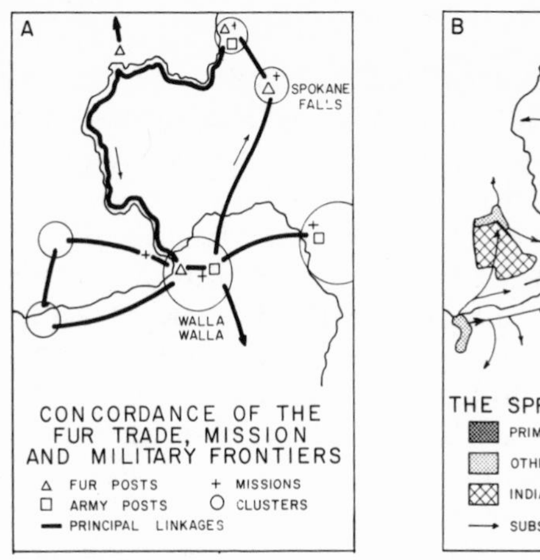

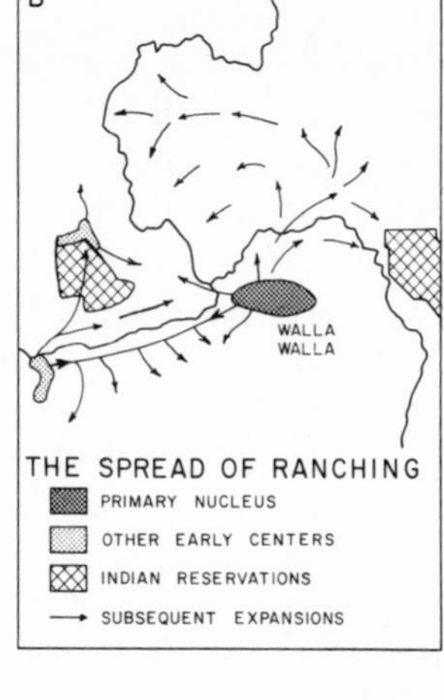

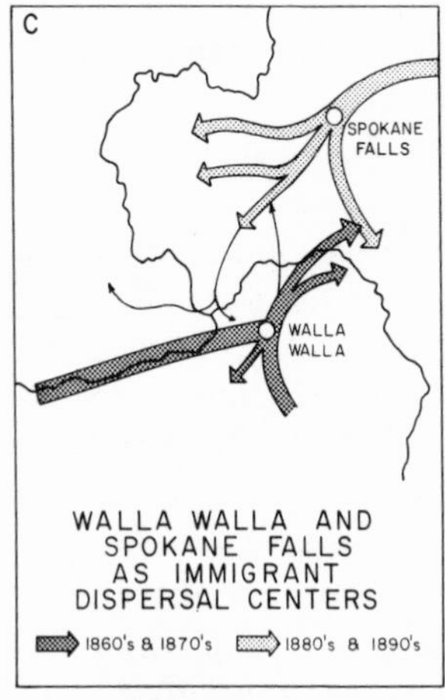

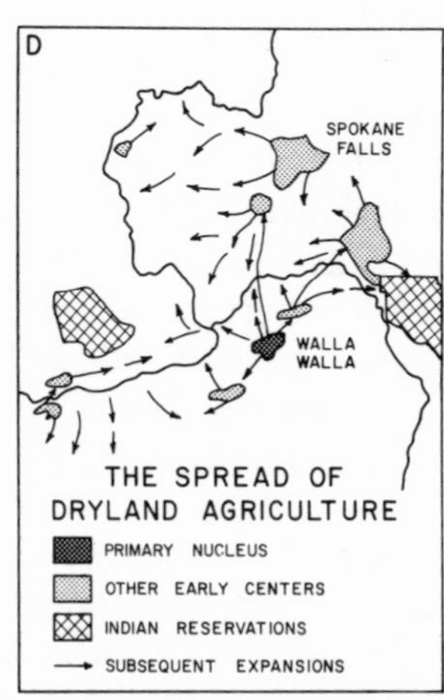

FIGURE 40 SETTLEMENT OF THE GREAT COLUMBIA PLAIN

Figure 40-A shows the first stage of settlement in which fur traders occupied a set of selected points already occupied by missions and army posts. Figure 40-B shows how ranching then came in to spread areally outward, primarily from a zone around Walla Walla. Figure 40-C shows the strongly linear impact of the railroads which served to establish a second dispersal center at Spokane Falls. Figure 40-D shows another areal expansion as dryland agriculture moved out from a number of points, including those established by the railroad.

After Donald W. Meinig, *The Great Columbia Plain: An Historical Geography 1805–1910* (Seattle: University of Washington Press, 1968), Map 48.

ways of our society. In their concern to integrate and understand the totality of cultural elements in an area, a number of cultural geographers have attempted to establish and clarify those ideas and institutions that exert the most effect on the character of landscapes. Following a fundamental overview in adjacent disciplines, Lowenthal and Prince have presented a series of essays that analyze how the perceptions and values of the people of England are related in their varied landscape. With a large body of detailed historical data available, it was possible to trace the evolution of the present landscape, and to record the way in which the English people preserved and enhanced it according to value systems that possessed considerable tenacity through the years. Similarly, in a symposium presented some years ago by the Association of American Geographers, scholars attempted to interpret the evolution of the landscape of southern California from the standpoint and within the value system of the local people. More recent work on the perception of the rural and urban landscapes has been undertaken by cultural geographers in close collaboration with psychologists. In a number of pilot studies Lowenthal has analyzed the responses of people who move along preselected routes in rural and urban areas that display a wide variety of features of different aesthetic quality. Through suitable multivariate scaling procedures, paths through these environments have been classified on the basis of essentially qualitative criteria.

Cultural geographers have frequently demonstrated their concern for the aesthetic deterioration of the American landscape, and have often been prominent in local-level attempts to correct the visual stress and imbalance billboard strips impose upon approaches to American towns. Numerous geographical contributions are published in the journal *Landscape,* which provides an important vehicle for the exchange of ideas between many disciplines and viewpoints that share a common concern for environmental quality and appearance.

In quite another setting, Sonnenfeld has worked with the Arctic Eskimo in an attempt to determine the degree to which his viewpoint of the environment differs from that of outsiders. Although such work, and that of other cultural geographers, is closely allied to research in environmental perception and behavior (see the section, "Environmental Behavior," below), it is representative of many studies that reveal connections between attitudes, aspirations, and the shaping of practice and policy. But it is clear in all these fields of investigation, that range from urban America to Arctic tundra, that more work is required before cultural geography can interpret people's preferences in urban and regional design, and before land and resource management can count on a working knowledge of cultural factors. As better links are forged to other disciplines such as engineering and psychology, the quality and quantity of geographic work on problems of this sort will improve rapidly. These problems are the natural domain of the cultural geographer because of his strong tradition of historical insight into the expansion and expression of systems of ideas as culture. For, regardless of his specialty, a cultural geographer would argue that the historical perspective and a grasp of all the varied elements that shape a landscape are essential for a sound understanding and evaluation of future policies that affect large numbers of people.

Urban Studies

Urban problems have long been a major area for geographic research. Urban studies were published by British and German geographers before the turn of the century, and by French, Swedish, and American geographers before World War I. Since World War II there has been an intensification of both research and educational activity in urban geography. Fully 40% of the dissertations published in the University of Chicago Research Papers since 1948 could be classified as urban, and, in the early 1960s approximately 20% of all geography dissertations were on urban topics. A survey of graduate chairmen in 1968 indicated that urban geography rated highest in growth among the substantive fields.

One of the handicaps to the development of urban-geographic work in the earlier postwar period was the lag in curricular development. Although there was much urban work at the graduate and advanced undergraduate levels, introductory courses showed the traditional emphasis on physical geography or on a world regional survey. The result was a sharp contrast between graduate training and initial teaching assignments. Although some British books were available for lower-level urban geography courses, it was not until the late 1950s and mid-1960s that textbooks incorporating some of the research findings on U.S. cities became available. Curricular development has been rapid during the 1960s, however, and urban work has become stronger at all levels. Geographers have actively participated in urban studies programs developed at major universities, and the High School Geography Curriculum Improvement Project, sponsored by the National Science Foundation through the Association of American Geographers, has emphasized urban study in the development of a new high school course, "Geography in an Urban Age."

The tie between urban geography and the models discussed in the locational analysis section is particularly close, although some of the concepts treated in cultural geography are also being considered in contemporary urban studies. The following discussion is organized according to a scale of observation. The first part deals with studies of groups or systems of cities; the second deals with the internal characteristics of the cities themselves; and the third deals with both varying internal and external characteristics of cities in different cultures.

Systems of Cities. The spatial organization of society is reflected in the arrangements of systems of cities as they have developed in regions, nations, and other areal units. Geographers have studied these urban patterns and the processes in their development at both the theoretical and the empirical level.

One important theme is the spatial pattern of urban places according to their size, functions, and the linkages among them. Regularities in the size and number of settlements in a region have been described by equations that relate urban population and urban rank. More recently, attempts have been made to incorporate these relationships into the framework of general systems theory. Central-place theory and its derivatives deal with patterns of urban size, spacing, function, and tributary areas. The original impetus for much of this work came from the study of southern Germany in 1933, by W. Christaller, who developed a general deductive theory to describe the size, number, and distribution of towns. (See the discussion in Chapter I, pp. 24–27.) Figure 33 (p. 57), shows an idealized system of cities and market areas that might be expected under certain assumptions. This simplified model served as the basis for a wide variety of empirical studies of the structure of urban systems, particularly in the United States and Western Europe. Figure 41 shows an example of an actual system of cities and tributary areas in Estonia. As the empirical studies progressed, both the assumptions and the implications of the Christaller model were subjected to

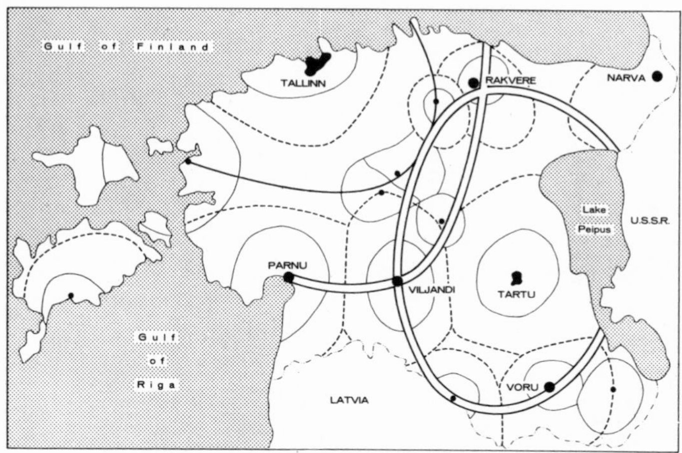

FIGURE 41 A HIERARCHICAL HINTERLAND SYSTEM

A hierarchical system in Esthonia. The lines enclose spheres of influence for different urban functions. The thick bands, for example, enclose tributary areas for the highest order functions, found only at Tallinn and Tartu. The broken and thin lines enclose tributary areas for lower-order functions.

After Edgar Kant, "Umland Studies and Sector Analysis," *Studies in Rural-Urban Interaction*, "Lund Studies in Geography, Series B, Human Geography, No. 3" (Lund: C. W. K. Gleerup, 1951), pp. 3–13, Fig. 1.

critical testing and modifications. Among the topics tested separately were spacing between cities of varying sizes, tendencies for centers to develop on divides of influence between larger centers, and comparative grouping of centers according to size or according to functions. A series of large bus-hinterland maps with much potential utility for planning was prepared for the Ordnance Survey in Great Britain in the 1950s that delimited a set of tributary areas. In the early 1960s, many of those empirical relationships between size, function, tributary area, spacing, and population density were formalized and expressed in both graphic and equation form by Brian Berry and his associates (Figure 42). Later theoretical work

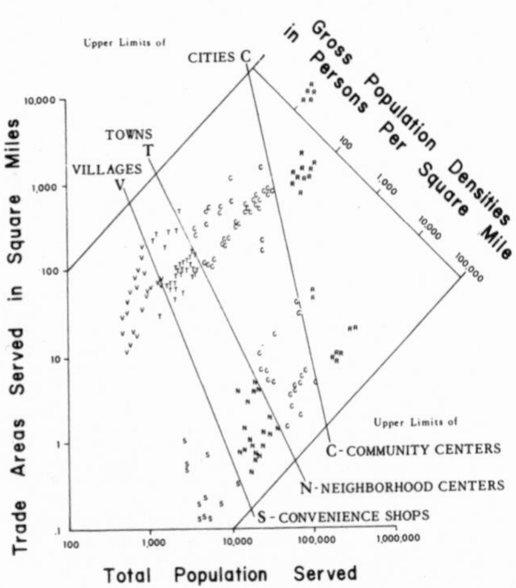

FIGURE 42 GRAPHS OF CITY SIZE AND HINTERLANDS

Some relations between central-place hierarchies, population served, and trade areas in regions of differing densities. Cities, towns, and villages in agricultural areas are shown in the upper part of the diagram. Different-size shopping centers in suburban and urban areas are shown in the lower part of the diagram.

From Brian J. L. Berry, *Geography of Market Centers and Retail Distribution* (Englewood Cliffs, N.J.: Prentice-Hall, Inc., 1967), Fig. 2.9.

moved in directions similar to those discussed in the section, "Locational Analysis": an attempt to incorporate observed regularities into a framework of general systems theory; a set of alternative geometric constructs related to the problem of packing different-shaped regions in a space; and attempts to develop explicit behavioral bases for the development of urban systems.

A second theme in the study of groups of cities has been spatial processes. Recent spatial-process studies have dealt with many topics—migration flows, diffusion of innovations, and the increased concentration of growth on the largest centers. Studies of migration flows and the diffusion of innovation have been affected by recent work on urban systems as well as by the work of Torsten Hagerstrand and other Swedish geographers on similar subjects. Figure

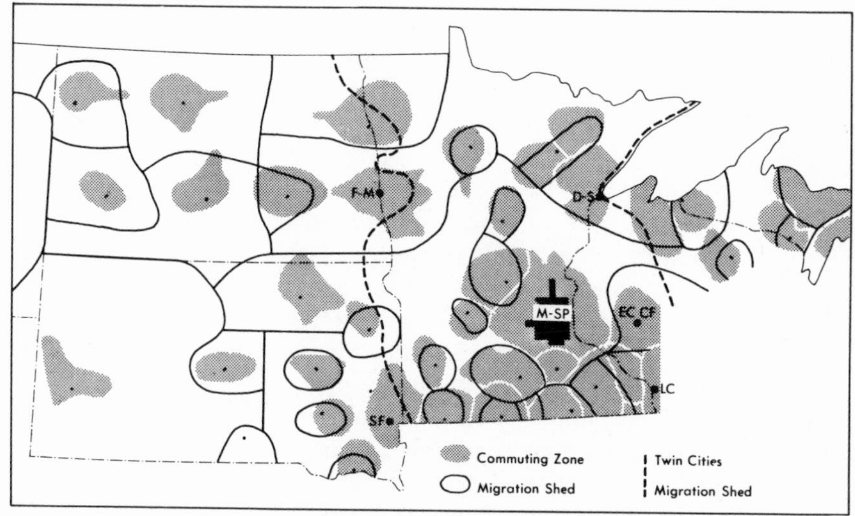

FIGURE 43 MIGRATION-SHEDS AND COMMUTATION RANGES: UPPER MIDWEST

Shaded areas represent the zone within which there is commuting to the indicated urban center. Lines indicating migration divides indicate zone within which more migration occurs to the indicated center than to any other center. M-SP—Minneapolis–St. Paul; LC—La Crosse; EC CF—Eau Claire, Chippewa Falls; D-S—Duluth, Superior; SF—Sioux Falls; F-M—Fargo, Moorehead.

After Russell B. Adams, *Population Mobility in the Upper Midwest,* "Urban Report No. 6, Upper Midwest Economic Study" (Minneapolis: University of Minnesota, May, 1964).

43 shows migration sheds, trade areas, and commutation ranges in the upper Midwest, an example of the relation between patterns of migration and the system of cities being considered. An understanding of the ways in which the cities are connected into a system is

clearly essential to an understanding of migration. The migration between towns, for example, is much heavier than the rural-to-urban flow; the different levels of the urban hierarchy are associated with a multiple-stage migration process; and commuting is becoming a substitute for migration as growing job centers reach out 30 miles or more to rural townships and small towns with inadequate local employment.

The Monte Carlo simulation models used by Torsten Hagerstrand and other Swedish geographers in their early study of migration and of the diffusion of innovation (see pp. 27–31) had considerable impact on U.S. geography during the late 1950s and early 1960s, and the models were applied to a variety of empirical studies including the diffusion of street railways among U.S. cities, the spread and growth of urban centers in a colonizing area, the evolution of central-place systems, inter- and intra-urban migration patterns, and the growth of transportation networks. As the empirical studies progressed, the relatively simple original model was modified correspondingly. The distance effect was broadened to include economic and social distance as well as distance through a multi level hierarchy; barrier effects were treated separately both in the abstract as reflective, absorbing, and permeable barriers, and in specific manifestations such as physical barriers, political boundaries, and social resistances. There was a growing interest in behavioral approaches to geography in the late 1960s, and studies of migration and diffusion were increasingly affected by the work of psychologists and sociologists studying similar problems.

The process of urban growth has received greater attention because of work that has dealt with the ways in which systems of cities have developed. Studies of the evolution through time of urban hierarchies have shown a tendency for centers to decline relative to centers at the next higher level in the system. Towns have tended to take over functions of villages, cities to take over functions of towns, and large metropolitan centers have tended to increase their dominance over surrounding systems of smaller centers. There is evidence that a similar process is at work among the metropolitan areas themselves. Figure 44 shows how the country's air-passenger traffic is dominated by a relatively few metropolitan centers. In the eastern zone, air-passenger traffic is dominated by New York or by cities such as Chicago, Atlanta, and others which, in turn, are dominated by New York. Cities in the western zone are dominated by Los Angeles and San Francisco. Between the two, a smaller zone focuses on Dallas and Houston. Studies through time indicate a

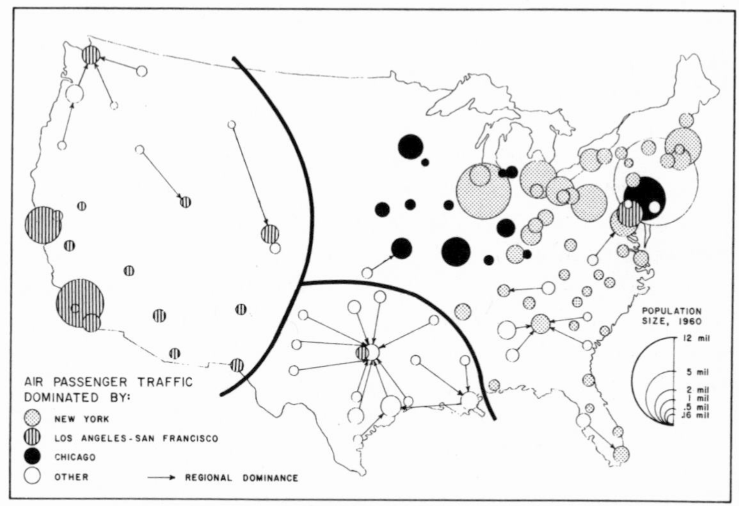

FIGURE 44 AIR-PASSENGER DOMINANCE

The heavy lines divide the country into three zones of air-passenger dominance. In the East is a zone within which the air passenger traffic of all cities is dominated by New York or, in a hierarchical structure, by cities such as Chicago, Atlanta, and others, which, in turn, are dominated by New York. The zone in the West encloses cities similarly dominated by Los Angeles, San Francisco, and Seattle. The hierarchy of dominance in the south-central zone focuses on Dallas, Houston, and New Orleans.

Courtesy Edward J. Taaffe, The Ohio State University.

small but remarkably consistent increase in the percentage share of New York and Los Angeles in the air traffic of all metropolitan areas.

Studies that show the increasing dominance of large centers have reinforced the "growth pole" ideas used by economists, planners, and geographers, in which it is postulated that economic growth does not take place evenly over area but rather tends to be concentrated at a particular center or group of centers that expand more rapidly than surrounding areas. As the spatial economy develops, these centers tend to accelerate and widen the gap between them and the surrounding areas. Figure 45 shows a series of profiles illus-

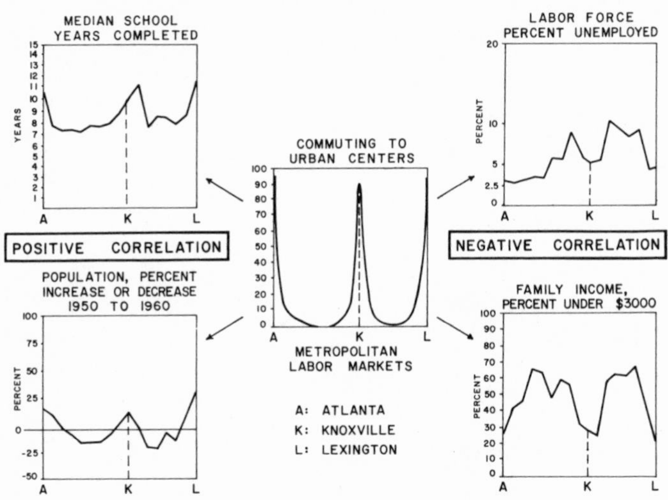

FIGURE 45 DEVELOPMENT PROFILES

Profiles of selected socioeconomic indicators from Atlanta to Knoxville and Lexington, showing the concentration of development on the largest centers. The three major centers are positively correlated with median school years completed and population increase, and negatively correlated with per cent unemployed and per cent having income under $3,000.

After Brian J. L. Berry, *Metropolitan Area Definition: A Re-evaluation of Concept and Statistical Practice*, Bureau of the Census Working Paper No. 28 (Washington, D.C.: Government Printing Office, 1968), Fig. 7.

trating this tendency for development to be concentrated on the largest centers. The graphs represent a traverse from Atlanta through Knoxville to Lexington, Ky. The three major metropolitan labor markets show a positive correlation with median school years completed and population increase, and a negative correlation with per cent unemployed, and per cent income under $3,000.

The Metropolis as a System. Studies of the internal spatial organization of metropolitan areas have been similar to the studies of groups of cities in that they have dealt with comparable themes and have been concerned with the processes leading to spatial patterns and, more recently, with behavior in space.

Spatial patterns within the metropolis have been studied in detail by geographers and sociologists. Early morphological studies in the

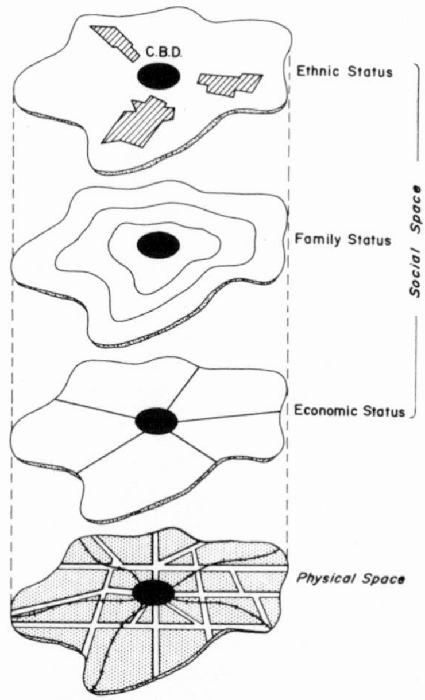

FIGURE 46 PATTERNS OF ECONOMIC, FAMILY, AND ETHNIC STATUS

An idealized spatial model of urban ecological structure and change. The model is altered from a circular form so as to demonstrate the effects of differential access along radial transportation routes. Economic status is distributed in radial sectors of high-, middle-, and low-income neighborhoods. Family status (type of household, fertility) varies concentrically from the center of the city. Ethnic status varies in clustered groups.

From Robert A. Murdie, *Factorial Ecology of Metropolitan Toronto, 1951–1961* (Chicago: University of Chicago, Department of Geography Research Paper No. 116, 1969), Fig. 2.

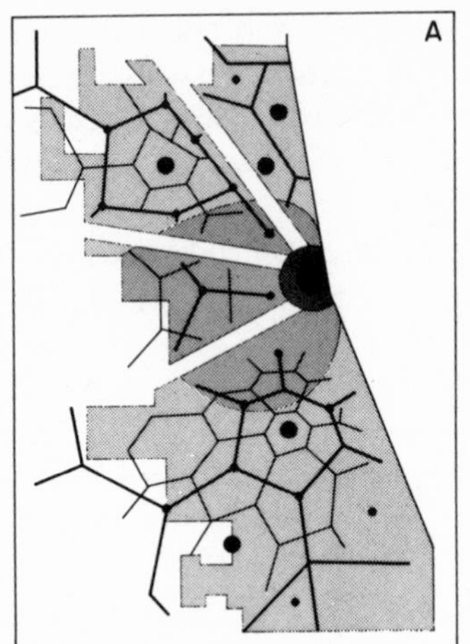

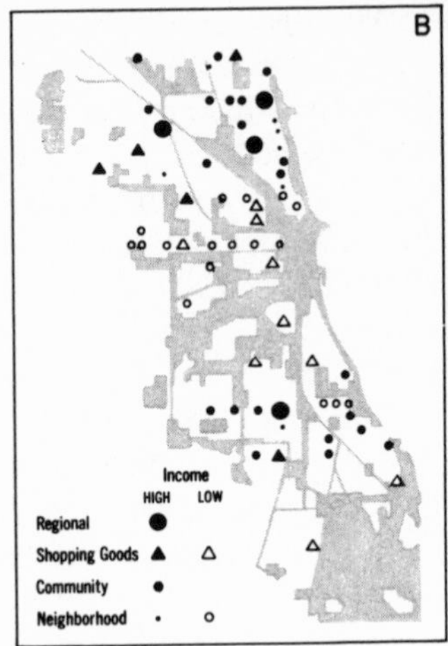

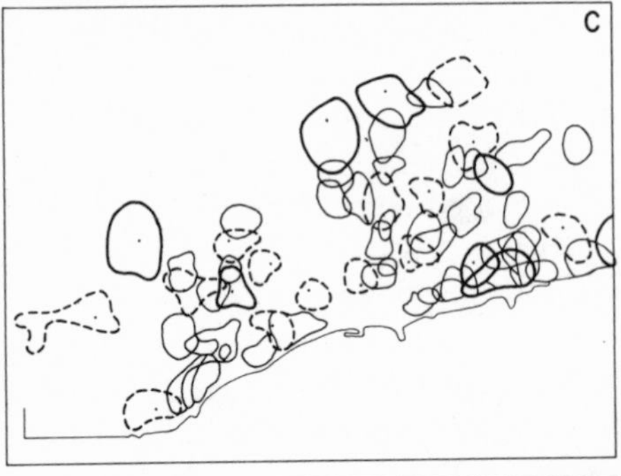

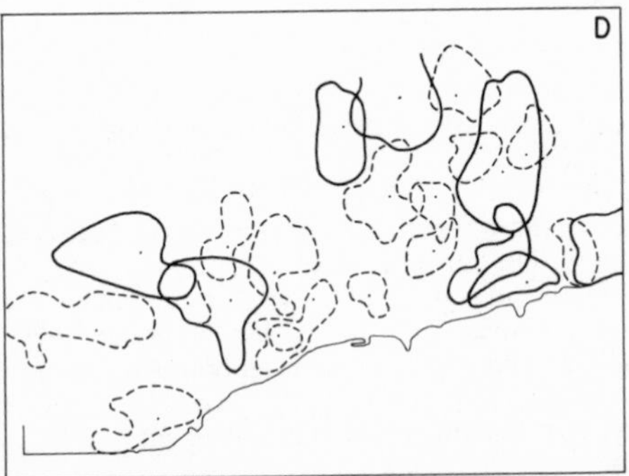

FIGURE 47 COMMERCIAL CENTERS AND TRIBUTARY AREAS: CHICAGO

Figure 47-A is an idealized representation of a system of trade centers and trade areas within a metropolitan area. Figure 47-B shows Chicago shopping centers classified into four levels, with

1920s and 1930s occasioned much debate over the relative merits of three ways of describing the pattern of a city's growth: the concentric-zone theory, the axial or sector theory, and the multiple-nuclei theory. Figure 46 represents the way in which some recent studies have indicated that the three models may be combined with some of the social-area dimensions developed by urban sociologists to form separate components of the total socioeconomic structure of city neighborhoods. Economic status seems to be distributed in sectors of high-, lower-, and middle-income neighborhoods radiating from the center of the city. Family status (fertility, type of income, etc.) tends generally to vary concentrically from the center of the city. Ethnic status tends to form clusters in different parts of the city. The study of population density and land values has resulted in the modification of distance-decay functions that have been rendered as two-dimensional surfaces, and related to both urban rent theory and information theory. The patterns formed by commercial centers and their hinterlands bear some relation to the idealized systems applied to the study of groups of cities, as is shown in Figures 47-A through 47-D. The departures from the idealized pattern are considerably more pronounced in the intraurban case, however, and alternative constructs have been used since the 1950s by urban and marketing geographers. For example, Figure 48 shows a classification scheme for types of commercial centers found within metropolitan areas in which the system had to be adjusted for planned and unplanned centers, different types of ribbon development, and clusters of specialized areas.

The process of urban growth has also been a subject of continuing interest in intraurban study. Historical studies of the growth of cities have been appearing in geographic journals since the turn of the century. Jean Gottman's study in the 1950s (in which he first identified the highly urbanized eastern seaboard area as "megalopolis") dealt with the tendency toward intermetropolitan coalescence

---

separate categories for low-income areas. The shaded areas represent industrial land. Figure 47-C represents tributary areas in Chicago for convenience goods; Figure 47-D represents tributary areas for shopping goods.

After James Simmons, *The Changing Pattern of Retail Location* (Chicago: University of Chicago, Department of Geography Research Paper, No. 92, 1964), cover design, Figs. 11, 13, and 14.

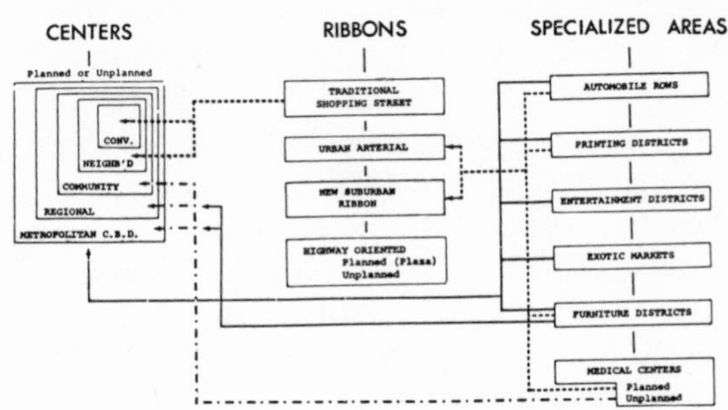

FIGURE 48 CLASSIFICATION OF URBAN COMMERCIAL AREAS

Typology of business areas within the metropolis. In addition to the business centers, which are organized in a hierarchical system, there are ribbon commercial developments and clusters of specialties.

From Brian J. L. Berry, *Geography of Market Centers and Retail Distribution* (Englewood Cliffs, N.J.: Prentice-Hall, Inc., 1967), Fig. 2.19.

as the suburbs of the major centers grew toward each other (Figures 49-A and 49-B). Recent developments in the study of urban growth processes have included careful sequential mapping and some initial attempts at computer simulation. Figure 50 shows two simplified versions of maps that were used in a study of Minneapolis in which the outward expansion of areas of high, medium, and low dwelling-unit density from 1900 through 1940 and 1956 was carefully mapped. The contours were then extrapolated to 1980 after correcting for such things as the tendency for the high-density areas to follow relatively level land and the medium density areas to follow traffic arteries.

Urban growth processes have also been studied for particular types of land use and for certain urban subsystems. Figure 51 depicts the spatial pattern of commercial centers undergoing drastic change as the metropolitan population continues to move outward. The resulting problem of dislocation for small business has recently been studied separately for a selected group of large cities. The Negro ghetto has been studied as a separate subsystem in the city both in

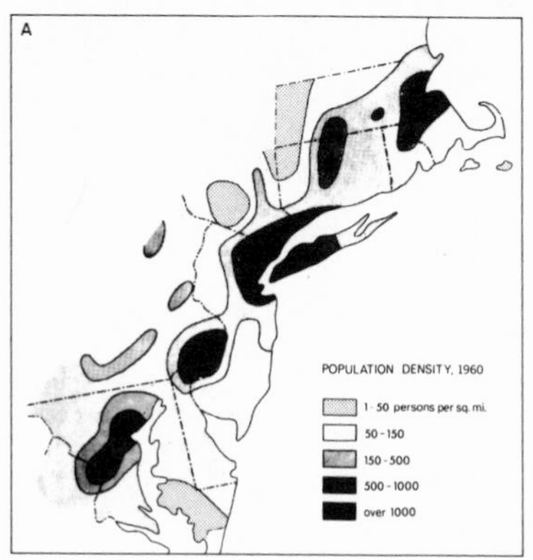

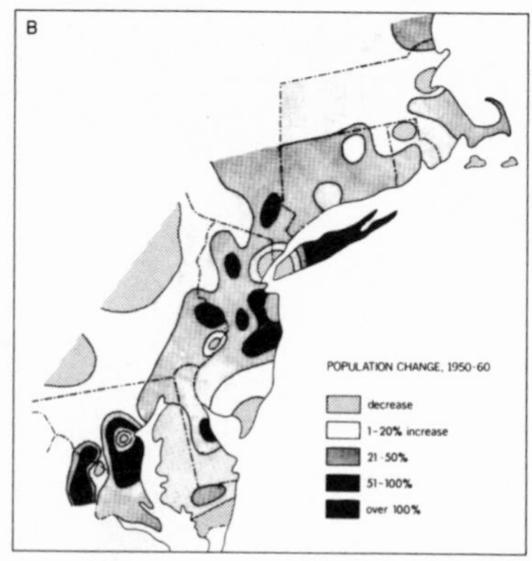

FIGURE 49 MEGALOPOLIS

Figure 49-A shows the 1960 population density in the Eastern seaboard by minor civil divisions. Densities of over 500 persons per square mile dominate much of the area between Washington and Boston. Figure 49-B shows the rates of increase by county from 1950 to 1960. The tendency toward intermetropolitan coalescence is evident in the high growth rates for the suburban counties between the central cities.

After Jean Gottman, *Megalopolis: The Urbanized Northeastern Seaboard of the United States* (New York: The Twentieth Century Fund, 1961), Figs. 1 and 9.

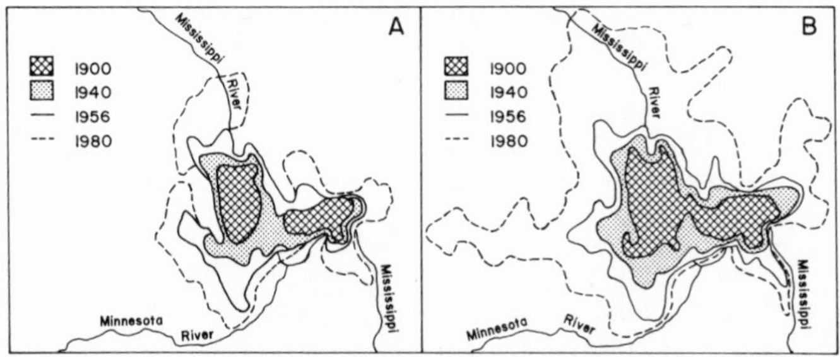

FIGURE 50 EXPANSION OF MINNEAPOLIS

Figure 50-A shows the outward expansion of areas of high dwelling-unit density in Minneapolis from 1900 extrapolated to 1980. Figure 50-B shows the expansion of the areas of medium dwelling-unit densities. There is a tendency for the high-density contours to follow level land and for the medium-density areas to follow traffic arteries.

After John R. Borchert, "The Twin Cities Urbanized Area: Past, Present, Future," *Geographical Review*, LI, No. 1 (January, 1961), 47–70, Plate II.

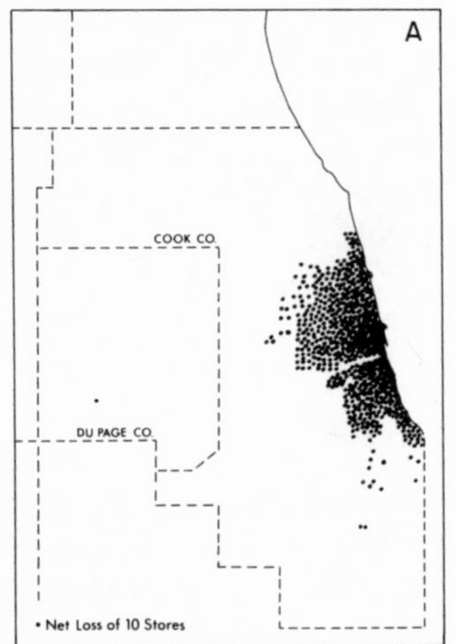

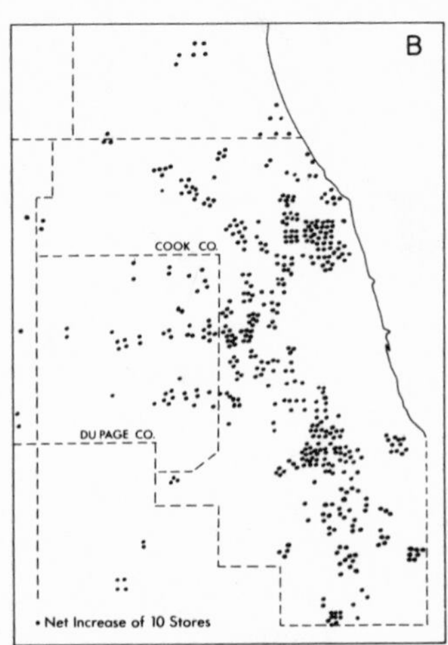

FIGURE 51 CHANGING COMMERCIAL PATTERNS: CHICAGO, 1948–1958

Figure 51-A shows the distribution of declining retail areas in Chicago. Each dot represents a net loss of ten stores. Figure 51-B shows the areas of increased retail activity. Each dot represents a net increase of ten stores.

After James Simmons, *The Changing Pattern of Retail Location* (Chicago: University of Chicago, Department of Geography Research Paper, No. 92, 1964), Fig. 27.

terms of its internal spatial patterns and its own growth process. Studies of commercial blight revealed significant differences both in hierarchical structure and in reaction to change in Negro commercial districts. Studies of the growth of the Negro ghetto have proceeded from the initial application of the Hagerstrand simulation model (see p. 31) to include the effects of rental levels, a "tipping-point" mechanism, and a birth-death demographic component. Studies being carried on by Harold Rose in Milwaukee suggest the possibility of a reasonable prediction of ghetto expansion.

The emphasis on behavior in the study of intraurban growth processes is also reflected in other recent studies that deal with intraurban migration, consumer-shopping behavior, and perceptions of neighborhood areas. Political behavior has also been receiving more attention from urban geographers. Voting behavior, for example, has been studied for the effects of acquaintance circles, city-to-suburb migration, spatial contagion, and group identifications at varying levels. An interesting example of a behavioral and political approach to the study of urban spatial organization is a study of the location decision-making process involved in urban redevelopment being carried on in Philadelphia by Julian Wolpert. Since most urban redevelopment, such as the building of a new expressway, involves a spatially unequal distribution of benefits as opposed to costs, certain groups will be detrimentally affected by proposed changes. Wolpert postulates the greater participation by potentially impacted groups in such decision making and suggests a sequence of game-theoretic structures as an initial step in the development of a theory of locational strategy.

The above examples represent only a few of the ways in which the research interests of the geographer have converged with those of the urban sociologist, engineer, city planner, regional scientist, and urban economist in studying the internal spatial organization of the city. Studies of the urban economy have prompted a common interest in city and regional input-output studies, multiplier effects, and locational patterns among geographers, urban economists, regional scientists, and planners; studies of urban transportation networks and related land-use patterns have been carried on by geographers, economists, engineers, and planners, and some of this interest is now focused in the attempt to develop land-use forecasting models for urban areas; central business districts, urban fringes, and individual types of land uses have also been investigated by geographers, urban sociologists, and city planners.

Cross-cultural Urban Studies. Comparative or cross-cultural urban studies represent a particularly promising field for expanded work as new approaches to urban geography begin to proliferate. As each of the ideas or models mentioned in this brief summary is applied to cities or systems of cities in different regions and cultures, it provides useful insights both into the nature of the model and its assumptions, and into the nature of the cultural context in question. Different value systems and different social organizations produce significantly differing spatial systems. In many instances the diversity of approach to these systems (see the section, "Cultural Geography") is combined with new approaches (see the section, "Locational Analysis") as shown in Figure 52, which illustrates how two differ-

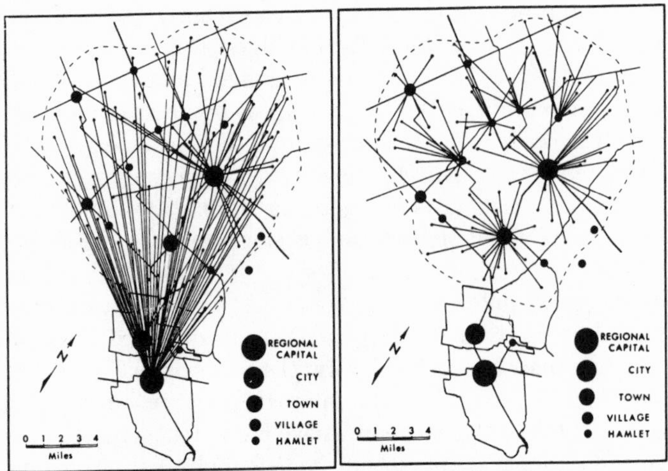

FIGURE 52 CULTURAL DIFFERENCES IN SHOPPING PATTERNS—MENNONITE AND MODERN CANADIAN

The map on the left represents the shopping travel of "modern" Canadians for women's clothing. The one on the right represents the shopping travel for yard goods of Mennonites residing in the same area. The Mennonites show considerably less dependence upon the large metropolitan centers.

From Robert A. Murdie, "Cultural Differences in Consumer Travel," *Economic Geography*, XLI (July 1965) 211–33, Figs. 14 and 15.

ent culture groups, even in the same region, may utilize the same set of central places in quite different fashions. The map on the left represents the shopping travel of "modern" Canadians for clothing;

the one on the right represents the shopping travel for yard goods of a group of Mennonites residing in the same area. The lesser dependence of the Mennonites on the large metropolitan centers is quite clear, and indicates the need for further modifications in central-place analysis. Applications of negative exponential distance-decay functions (see Chapter I, pp. 9–13) for population density in Asian and Latin American cities reveal a number of interesting variations on the basic pattern. The growth-pole idea has been applied in underdeveloped areas—for example, in Kanpur, India—to point out the need to increase the "trickle-down" effect from the metropolitan center into the surrounding rural hinterland with its smaller towns and villages. Paul Wheatley has recently completed a study of the origins of Chinese cities, which is a significant contribution to the social scientific literature on the origins and development of all cities on a cross-cultural and cross-regional basis. Other studies have noted variations in urban hierarchies, in intraurban segregation patterns, city-hinterland relationships, and social area and factorial ecology patterns. Further studies by urban and regional geographers are needed to provide better understanding of the effects of cultural differentiation on the range and nature of variation in the process of urbanization and on the applicability of Western technology in different parts of the world.

Environmental and Spatial Behavior

One of the new research directions rephrases the question of the traditional concern of geographers for man-environment relationships in terms more responsive to the application of modern social science research techniques. Changes in standards of value and issues raised by the development of technology have led to a substantial growth of interest in environment. Geography is one of the social sciences converging on this problem.

The specific interests of geographers have focused upon two main themes: the natural environment, including elements of weather and climate such as floods, droughts, and earthquakes; and spatial and physical characteristics, especially of cities. The approach adopted is largely behavioral. The environment is viewed not as something fixed or as imposing limits but rather as a function of culture and technology. Everything is subject to change—even the poverty of man in the tropics—by the application of technology to the environment and by the process of adapting human society and known values to take advantage of the opportunities created.

The new research direction has followed in the tradition of Harlan H. Barrows, who viewed geography as human ecology and who made important contributions to federal resource policy and planning in the 1920s and 1930s with the National Resources Planning Board and the Tennessee Valley Authority. This concern for public policy issues led to a concentration on flood problems; this was led by Gilbert F. White at the University of Chicago, and supported by Robert W. Kates and others. The success of the research efforts has had a dramatic impact upon the thinking of geographers about environmental problems. Research scholars and published research results are still numerically small, but they have concentrated on the concept of environmental perception, briefly discussed above in "Geography and the Social Sciences," and are demonstrating that the ways in which men structure their views of the external world in their own minds help to explain patterns of decision-making behavior that have hitherto been poorly understood or simply dismissed as irrational.

Environmental Perception. Among the studies that deal explicitly with the physical environment, the floodplain research provides an example of the concentration of effort on a single, well-defined problem, namely the meaning and significance of one environmental element to the residents and managers of floodplain property, and the role of perception in decision-making. The main point of the work has been to place the problem of floods in a broader perspective than that of the traditional engineering approaches. This broader framework consists of examining all the possible human adjustments to floods and systematically searching for the reasons why some alternatives have been given greater stress than others in the past. The framework also systematically examines the limits and conditions of largely untried solutions. Engineering adjustments for flood control have been stressed by government agencies, for example, while, until recently, relatively little work has been done on social adjustments such as floodplain zoning, insurance programs, and flood-proofing of buildings.

Although a wide range of choice exists both for government agencies and for floodplain residents and managers, the choice is often greatly constrained in practice. In the past it has often amounted to no more than an apparent choice for residents to suffer flood losses or to press the government for action to control floods by engineering structures. For government agencies, the choice has often been to recommend construction of a dam, or to do nothing. Obstacles to the consideration of a wider range of alternatives have been identified and in some cases removed.

The decision-maker's choice is affected by his perception of the flood hazard. The likelihood that a flood threat will be recognized and that adjustments will be adopted is related to flood frequency. The relationship is even stronger where actual frequency is replaced by perceived frequency. That some individuals living in similar situations perceive flood threats and others do not is not well accounted for by differences in education or income. There is, however, a relatively consistent and predictable relationship between hazard perception and experienced flood frequency. Flood frequency is not viewed as continuously variable but rather as a discrete distribution between negative certainty, positive certainty, and an intermediate zone of relative uncertainty (see Figure 23).

Further study has indicated some significant differences between the hazard perception of the floodplain dweller and the perception of the technician-scientist. The floodplain resident or manager tends to impute a regularity to the occurrence of floods, and seeks to remove the uncertainty inherent in the situation by making it determinate and knowable. The technician-scientist, however, shows a much greater tolerance of uncertainty. One result of these differing perceptions is that floodplain residents and managers may acquire a false sense of security in the period after "the hundred-year flood," or after the construction of a flood-control dam. This, in turn, affects their receptivity to proposals for alternative damage-reducing adjustments and may lead to a relaxation of preparedness to deal with emergency flood threats.

Comparisons may also be made between different hazard situations. For example, interviews with floodplain dwellers from sites with varied physical situations and land use have been compared with responses from residents of the outer coastal areas of the eastern seaboard. There is clearly a higher order of hazard perception and adjustment adoption among coastal respondents than among the floodplain dwellers, as shown in Figure 53. The reasons for these

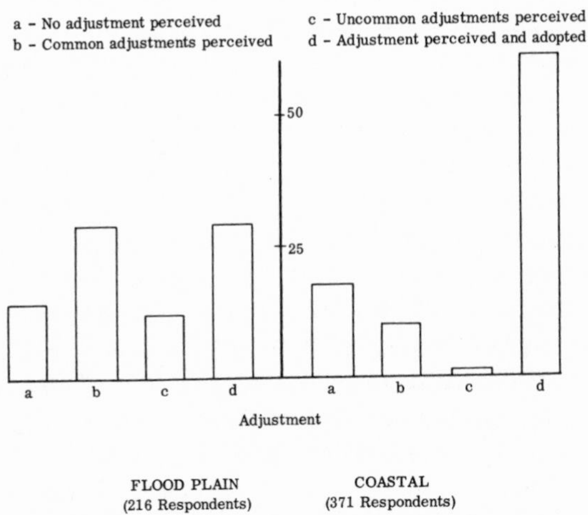

FIGURE 53 COMPARATIVE HAZARD PERCEPTION: FLOOD PLAIN AND COASTAL DWELLERS

The bar chart comparing the adjustments of flood plain and coastal dwellers to flood hazards shows a clear difference. More than 50% of the coastal dwellers interviewed had perceived and adopted adjustments as compared to a little more than 25% of the flood plain dwellers.

From Ian Burton, Robert Kates, Rodman Snead, *The Human Ecology of Coastal Flood Hazard in Megalopolis* (Chicago: University of Chicago Research Paper No. 115, 1969), Fig. 50.

differences could be elaborated at some length, but it seems clear that the high degree of hazard perception by coastal residents leads to a high rate of adoption of alternative adjustments. This finding suggests that the public information and education programs being developed by the U.S. Geological Survey and the U.S. Army Corps of Engineers might yield more returns on the coast than on the interior flood plains. Paradoxically, it seems to be more worthwhile to invest in such programs where knowledge of hazard is already at a high level and is followed by adoptions, than where knowledge is low and more information does not significantly increase the adoption rate.

Tangible results of these and related studies at the federal level have included the development of new planning philosophies under which a wider range of alternatives can be considered—as described in the NAS-NRC report, *Alternatives in Water Management,* and as exemplified in the report of the President's Task Force on Federal Flood Control Policy.

This kind of work is now being extended to other natural hazards in North America as well as to other nonwestern culture areas. Saarinen has explored the relationship between personality characteristics as shown in projective tests and perception of drought hazard in the Great Plains. The notion that personality types may respond to environment in radically different ways is also being explored by Joseph Sonnenfeld by means of color-slide tests and semantic differential tests in Alaska and Delaware. In a study of the recreational use of water-supply reservoirs, it has been found that activities are significantly more restricted in the Northeast and the Far West than in the remainder of the United States. An explanation of this contrast is found in the level of knowledge of disease and water-treatment technology prevailing at the time when surface water supplies were developed in various parts of the country. For example, attitudes and perceptions developed in New England in the prechlorination era set the emphasis on measures to protect the purity of upland supplies of surface water, and these attitudes still persist, in spite of overwhelming scientific evidence that the measures are no longer helpful and comprise an obstacle to the expansion of recreational opportunities.

There has been a growing awareness in the geographic literature of the political context in which resource management decisions are being made. This context is stressed in portions of *The Science of Geography,* in which problems such as the discontinuity between the administrative machinery of the federal-political system of the United States and the spatial phenomena to which they are related were suggested as significant research themes. Attempts have been made by Ackerman and Lof to codify norms for these relations in the light of modern technology. In another vein, Michel's study of the Indus basin and its transformation through time under a number of regimes stands as a contribution to environmental-perception and resource-management literature as well as a study in political and regional geography.

Geographic research in the problems of resource management has become greatly concerned with the social consequences of dramatic advances in technology and the application of this new knowledge to the control and modification of environment without corresponding steps to adjust human behavior and institutional responses to the new situation. Geographers have been contributing to studies of the consequences of weather modification by a National Science Foundation Special Commission, a Task Group on Human Dimensions of the Atmosphere, and in other ways. A recent survey of the geographic literature shows the tentative and preliminary nature of most of the earlier work on the human response to weather. As the cost of environmental information services increases, it becomes more important to know how people use weather forecasts and in what ways the information can be gathered and disseminated to maximize its utility and effectiveness. The dramatic, and international, correlation between the distribution of tropical climates and poverty clearly emphasizes the need for improved understanding of the role of climate in economic development and the economic consequences of disease. Major research efforts with contributions from geographers as well as other specialists are needed in these areas. Unfortunately, at present the manpower shortage has slowed progress in these areas, although it is widely recognized as an urgent need. Geographers are also working with the UNESCO Commissions on the Arid Zone and the Humid Tropics, and have established a Commission on Man and Environment as part of the work of the International Geographical Union.

Spatial Perception. Research on environmental quality, recreational amenities, and the urban landscape focuses on man's awareness of his environment and his view of the spatial organization around him.

Lucas has shown the various ways in which different users of a recreational area perceive the environment (see p. 33). In a study of the Meramec River Basin, Edward Ullman has fitted curves for high-, low-, and middle-income groups. These curves were then used to predict attendance at alternative recreational reservoirs before choosing which reservoirs to build (Figure 54). This method can

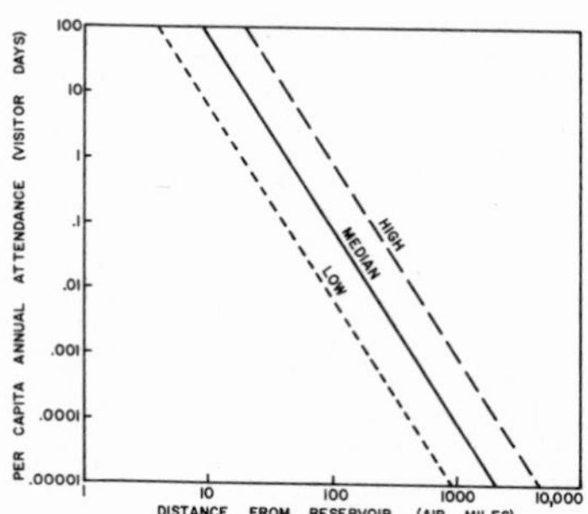

FIGURE 54 ATTENDANCE AT RECREATIONAL RESERVOIRS

Graph showing relationship between per-capita attendance and distance from St. Louis for selected reservoirs. The three lines show how this relationship varies for visitors of high-, medium-, and low-income groups.

From Edward L. Ullman, "Geographical Prediction and Theory: The Measure of Recreation Benefits in the Meramec Basin." In Saul B. Cohen, ed., *Problems and Trends in American Geography,* Chapter 10 (New York: Basic Books, Inc., Publishers, 1967), pp. 124–45, Fig. 10-1.

be refined to take account of the differing perceptions and space preferences among types of reservoir users (Figure 55).

Quite a different sort of environmental quality is involved in urban studies that focus on the individual's perception of stress in his environment. The decision to move or remain in a particular neighborhood is a function of perceived stress. A preliminary ecological system model has been developed that provides insight into the nature and degree of stress that causes urban dwellers to change neighborhoods. The potential value of such models to help modify the social and economic problems of neighborhood change is very great.

In addition to investigating the perception of environmental quality, hazard, and stress, geographers have devoted attention to the perception of geographic space in terms of over-all residential desirability. Starting from a number of pilot studies in the United States (see p. 35), residential perception surfaces have been constructed for a number of other countries and regions. In Great

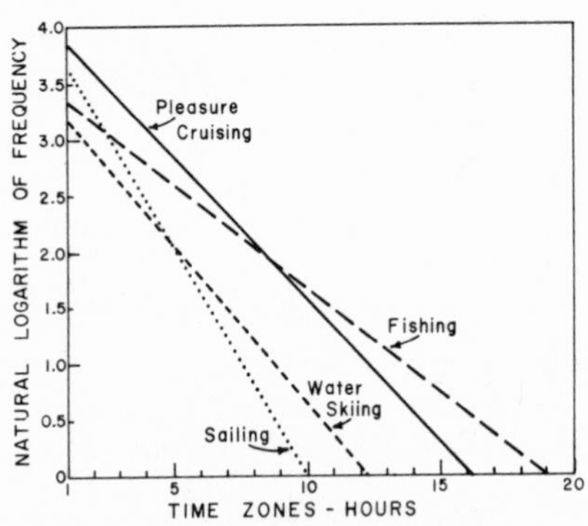

FIGURE 55 DIFFERING SPACE PREFERENCES BY USERS

The relation between number of participants and hours of travel for different recreational activities in a survey of fifteen Ohio lakes. Different sensitivities to distance are clear for different types of activities. The number of people sailing or water-skiing drops off sharply as travel time increases, while the number fishing seems to be least sensitive to distance.

After Barry Lentnek, Carlton S. Van Doren, and James R. Trail, "Spatial Behavior in Recreational Boating," *Journal of Leisure Research*, I, No. 2 (Spring, 1969), 103–24, Fig. 6.

Britain, for example, perception surfaces were constructed for a large number of pupils who were about to leave school and look for their first jobs. Multivariate and cartographic analysis indicated a high degree of regularity and agreement about the residential desirability of various areas. Samples from many schools, ranging from southern England to northern Scotland, indicated that the perception surface of the general viewpoint of children at a particular location could be broken down into a general national trend upon which a peak of local desirability was distorted in varying degrees. The degree of distortion between the general surface and the local effect varied with the location of the perception point in a regular and predictable manner. The way in which perception surfaces change with the age of the respondents has been investigated extensively in the western region of Nigeria. By subtracting a perception surface of twelve-year olds from one describing the viewpoints of adults in their early twenties, it was possible to map and isolate the changes in perception between those age groups. Strong local effects upon the younger group faded as the groups approached adulthood, and their residential images were reconstituted along the major urban and transportation axis of the western region, from which flows of information are far stronger than those of the relatively undeveloped, outlying rural areas.

Within urban areas there has been considerable interdisciplinary work between geographers, urban sociologists, and psychologists. A historical analysis of residential changes and purchases in the Twin Cities area indicated a clear directional bias among new purchasers to relocate on an axis that joins the previous residence with the center of the city. Similarly, urban sociologists in Holland have investigated the way in which the perception of "gateways" to the central business district changes with the location of the perceiver. Psychologists are also beginning to move from highly structured experiments in the laboratory to more freely structured inquiries regarding the perception of distance in geographic space. In conjunction with geographers in the United States, Stea has raised a number of questions about the strength of goals in geographic space and the perceived barriers to movement; while in the United Kingdom, the pioneering work of Lee has examined the way in which the tractability of young children varies in a regular manner with the distance of the journey to school. Additional work is being undertaken by geographers on the perception of distance to prominent world cities through analysis of error matrices. These distinct errors in perception have also been used in this study of locational ability based on the varying information content of a succession of maps. A recent compilation of work on perception surfaces and mental maps by 60 researchers indicates the rapid growth of interest in this interdisciplinary area.

The behavioral approach to urban environment is well exemplified by Lowenthal's work. Various groups of students followed carefully selected walks through unfamiliar parts of Cambridge, Massachusetts, and then responded to semantic differential and other tests about their impressions of what they had experienced. Preliminary results indicate sharp contrasts in the appreciation of the same environments between students of landscape architecture and those in English literature. These parallel the differences in flood-hazard perception between technician-scientists and laymen, as described earlier.

The growth of work in environmental and spatial behavior is a promising new development in geography. Common interests with other social scientists are fully apparent in this field. A recent issue of *The Journal of Social Issues* has been edited jointly by a geographer and a psychologist. The papers in this issue on man's response to the physical environment and those in *Environmental Perception and Behavior* edited by Lowenthal demonstrate the breath and the essentially interdisciplinary nature of this research. A journal, *Environment and Behavior,* has recently been established to deal with the study, design, and control of the physical environment and its interaction with human behavioral systems. Geographers, together with psychologists, planners, and others, are serving as associate editors and members of the editorial advisory board of the journal.

At the same time there is a closer relationship between the various research frontiers in geography; studies of environmental perception are making more use of some of the models discussed in "Locational Analysis." As discussed earlier, these models are, in turn, being revised to incorporate behavioral variables.

Research into the behavior of peoples and societies in relation to environment is being expanded vigorously in geography. There is a renewed sense of the magnitude and urgency of the problems, and a renewed sense of the contribution of geographic research to their solution. At the same time, a more sober attitude prevails in the wake of a new recognition that the problem of environment is fraught with dangers and is not readily susceptible to manipulation by techniques and methods that have succeeded in the past. There is an acute need for researchers trained in the new techniques that are demanded for the solution of these problems.

GEOGRAPHY AND PUBLIC POLICY

Geography in the United States has had a long tradition of public service and forming public policy. After World War I, Isaiah Bowman and other geographers advised the American delegation at Versailles. Others were employed in military agencies and in civilian agencies such as the War Shipping Board, where their knowledge of the commercial geography of the time was particularly valuable. Geographers were also employed in a number of government agencies during the 1920s, and cast their mark most heavily in the Soil Conservation Service; the Department of State, which maintained an Office of the Geographer; and the Department of Agriculture, as illustrated by the contributions of Oliver E. Baker and later Francis J. Marschner.

The cataclysmic events of the 1930s brought a new level of participation in national governmental policies. In those years major emphasis was focused on regional planning and development as the federal government became increasingly concerned with the economic affairs and social welfare of the country. Geographers played major roles in producing a series of policy-oriented publications, of which *Regional Factors in National Planning* (1935) is perhaps the best example. In addition, new emphasis came to be placed upon water resources and upon multipurpose river-basin development as an aspect of regional planning. Strongly influential in the development of water-resources policies were H. H. Barrows and his students, and geographers influenced the development of land use and resource survey and resource-management policies in the Tennessee Valley Authority. In absolute numbers the men involved were relatively few, but their influence was substantial in many sectors of American life.

The outbreak of war in 1939 and the participation of the United States in a global conflict marked a second major event in the recent history of geographers' direct involvement in government affairs. The universities were virtually drained of staff and students as geographers migrated to Washington, D.C. There, and later overseas, they found their niches in the Office of Strategic Services, which at one time employed more geographers than any other institution in the country, in other agencies in the defense establishment, and in the older civilian government agencies, the responsibilities of which had been greatly modified by the necessities of a war economy. It was during this period that geographical expertise in foreign areas proved particularly valuable, but at the same time inadequacies in training and in field experience exposed the necessity for developing training programs that would supply regionally informed specialists both in geography and in the other social sciences. It was recognized that even greater knowledge of foreign areas and of cross-regional and cross-cultural studies was crucial for the national welfare.

In the post–World War II period, and particularly after 1950, the role of geographers in the federal establishment began to assume its present form. Although many continued to hold research positions in the intelligence and defense agencies, others began to apply their knowledge to area-development problems in the Department of Commerce, regional agricultural problems in the Department of Agriculture, problems of policy in the Department of State, aspects of land use and water resources in the Department of the Interior, reference research in the Library of Congress and the National Archives, the Legislative Reference Service of Congress, and gathering and processing census materials in the Geography Division of the Bureau of the Census.

Geographers in the Geography Division of the Census are responsible for a new series of distributional maps of the United States, which illustrate major aspects of the spatial organization of contemporary American society. Agricultural policy is influenced directly by the work of the geography staff and advisers to the Secretary of Agriculture. Area- and regional-development policies are conceived and applied by geographers and other social scientists in the Economic Development Administration of the Department of Commerce. Much of the recent federal interest in outdoor recreation and in the preservation of open lands has been implemented by geographers working in the Department of the Interior. The U.S. Geological Survey has recognized the significance of geographical contributions to its work by establishing the post of Chief Geographer, and the National Atlas of the United States, now in production, is one of the Chief Geographer's responsibilities. It is estimated that there are 400 geographers in the federal establishment, or nearly 10% of the membership of the Association of American Geographers, a relatively high proportion for a social science discipline.

Not all contributions come, however, from geographers in national government employment. Equally significant are the effects on policy of geographers within the academic community. For example, in water resources management the work of Gilbert White and his students has had a profound effect not only through the President's Water Resources Commission and the Committee on Water of the Earth Science Division, NAS-NRC (which White chairs), but also through other agencies including the Corps of Engineers and through the traditional medium of academic publications. Similar policy developments in outdoor recreation, highway and other transportation expansion, and metropolitan area definition and planning, have been influenced either indirectly by the research of geographers on these topics, or more directly by their membership on committees of the National Research Councils in which geographers are currently serving in approximately 75 committee and other advisory positions. Government agencies are also drawing upon the intellectual resources of the universities, either by enlisting geographers as advisers or by granting contracts, as the Bureau of Public Roads and the Department of Housing and Urban Development have done for research on urban structure and metropolitan spatial relations.

At both local and regional levels geography has made considerable contributions, of both personnel and concepts. A high proportion of graduates of geography training programs in major universities find employment in city, state, and regional planning agencies. The former Directors of Research in both the Philadelphia and Chicago Planning Commissions were professional geographers, as was the President of the Washington Center for Metropolitan Studies. In the late 1960s there were more geographers than workers in any other field on the staff of the Los Angeles Planning Commission. Geographers play major roles as consultants in the formulation of policies on metropolitan planning and administration, and they do research on the causes underlying the growth and decline of neighborhoods and the transformation of retail marketing patterns within cities. A current study at the University of Illinois in Chicago, for example, is designed to develop and use a set of urban-growth models that can accept alternative policies as inputs and, through simulation or other means, evaluate the probable consequences of these policies upon the city's growth and development.

At the regional level, geographers have been working with economists and engineers in major projects such as the Upper Midwest Study, in different aspects of the studies of Appalachia, in the Northeast Corridor transportation study, and many others. Work with state agencies has also been expanding, and a number of excellent state atlases and state-level studies have been produced. The recently published *Geography of New York State* is an example of a work that combines information of potential practical utility to state agencies with a conceptual discussion of an area.

Geographers' contributions are also substantial on the international scale. The Lower Mekong Basin Project owes much of its direction to the development of policies that geographers have formulated in collaboration with other social scientists. The United Nations employs geographers as consultants and as members of their technical-assistance missions on problems of economic development and metropolitan and regional planning. The AID-financed Southeast Asian Development Advisory Group of the Asia Society includes a number of geographers (particularly in urban development), and the Asian Development Bank has recently initiated a major regional transportation study with the aid of a senior geographical consultant. Some interesting work is being done by the NAS-NRC Committee on the Development of Water Resources in Africa, chaired by Gilbert White. The research of T. R. Lee has contributed to the formulation of policy for community water supply in developing countries. His studies in India were part of the geographical contribution to the work of the Calcutta Metropolitan Planning Organization supported by the Ford Foundation. Moreover, the knowledge geographers have been acquiring—and are capable of expanding—that concerns resource endowments, urban development, regional transportation systems, settlement hierarchies, and local and regional ecosystems is beginning to make its mark on the policies of countries collectively described as "underdeveloped," as well as on the more prosperous ones.

The increased involvement of geographers in the efforts of behavioral and social scientists to seek better solutions to policy and planning decisions is due to the fact that there is a locational attribute to almost every such decision. The geographer brings to such decisions a concern for the effective management of space and a concomitant concern for the improvement of environmental quality. He also brings a set of techniques such as cartographic analysis, remote sensing, and the use of a large variety of spatial models, all designed to provide better insight into the nature of difficult locational questions. There is need for an intensification of geographic involvement in interdisciplinary policy formulation and implementation on at least three currently active fronts: urban problems, regional development both domestic and foreign, and problems of environmental quality. The concern of geographers with spatial organization and the ecosystemic problems of society, and the ready application of their approaches to the problems of the real world promise contributions of great value.

3
STATUS, TRENDS, AND NEEDS OF GEOGRAPHY

THE UNITED STATES AND THE WORLD

Geography in the United States has had a strong international component from its inception, as witnessed by the examples of foreign research cited in this report. The methodological roots of contemporary geographic research may be traced to nineteenth-century British, German, and, to a lesser extent, French geography, and ties between the United States and these countries have continued to the present. Ties with British geography have remained particularly strong. The relatively high proportion of foreign graduate students studying geography in the United States comes largely from the British Commonwealth, and there are many British faculty members at major United States universities. Since World War II, relations with Swedish and Canadian geographers have grown in importance and there have been increasing liaisons with geographers in the U.S.S.R., Poland, and Japan. Relations with the Canadians are particularly close. The president of the Association of American Geographers in 1969 was a Canadian; Canadians participate in many A.A.G. committees, and there is much two-way movement of faculty and students across the border.

The International Geographical Union, with 50 to 60 member countries, provides an institutional framework for cooperation among geographers and holds international congresses every four years. The first international geographical congress was held in Antwerp in 1871, although the I.G.U. itself was not organized until 1922. One important function of the I.G.U. has been its many commissions, each devoted to a topic of specialized concern in geography.

Recent listings in *Orbis Geographicus*, an international directory of geographers, included 81 countries in 1960. The United States has the largest number of *Orbis* listings, with Great Britian second and West Germany third. The listing, however, is by no means a complete or accurate picture of either absolute or relative magnitudes. There are only 74 listings for the Soviet Union in 1960, for example, although the professional association of Soviet geographers had 15,000 members in 1964. Neither do the listings provide an accurate picture of the relative importance of geography in each country. Geography in Britain, for example, plays a more conspicuous role in the university and secondary school curriculum than it does in the United States. In Canada, geography has undergone a rapid recent expansion that has tripled both faculty and graduate enrollment during the past five years.

The nature of geographic research also varies in emphasis among countries. Most other countries place greater stress on physical geography than does the United States. In addition, British geography shows a more strongly developed historical geography. In a number of countries there is an emphasis on applied work. To illustrate, *Orbis* lists 418 institutes and 603 government agencies engaged in work that might be classed as geographic. This is particularly true in countries such as Sweden, Poland, the U.S.S.R., and Canada where governments are actively engaged in planning and regional development.

The United States occupies a prominent place in world geography, however. Together with Sweden it has been a leader in mathematical work, and recent work in urban geography, locational analysis, and environmental perception has received major impetus from developments in the United States.

In the remainder of this report the focus will be on selected trends, characteristics, and needs of geography in the United States. The information is based primarily on five types of sources: (1) a questionnaire distributed in early 1968 by the Behavioral and Social Sciences Survey Central Planning Committee to Ph.D.-granting departments within universities (which we will refer to as the departmental questionnaire or, simply, the Survey);* (2) academic-degree

* The questionnaire survey is described more fully in the appendix to the general report of the Central Planning Committee, *The Behavioral and Social Sciences: Outlook and Needs* (Englewood Cliffs, N.J.: Prentice-Hall, 1969).

projections made by the Survey Committee; (3) sources of manpower, research, and educational information published by the Office of Education, the National Academy of Sciences, and others; (4) special studies of geography published by the Association of American Geographers and the Commission on College Geography; and (5) some separate unpublished surveys conducted by the Association of American Geographers.

MANPOWER

Survey data from questionnaires and other sources that deal with geographic manpower suggest that the most rapid increases in the last five years relative to other behavioral and social science disciplines have been in the number of bachelor's degrees awarded and the professional association memberships; the most conspicuous lag has been in the number of doctoral degrees. Other manpower indicators such as the size of faculty and the number of graduate students show growth roughly comparable to behavioral and social sciences averages. The net effect of these trends is to make clear a

TABLE 1 MEMBERSHIP IN THE ASSOCIATION OF AMERICAN GEOGRAPHERS, 1957–1968

Year	Geography Membership	Per Cent of 1957
1957	1,650	100
1958	1,701	103
1959	1,724[a]	104
1960	1,746	105
1961	1,810[a]	110
1962	1,873	114
1963	2,132	129
1964	2,958	179
1965	3,456	209
1966	3,439	208
1967	4,414	268
1968	5,570	338

Source: Association of American Geographers.
[a] Estimated.

serious and continuing manpower shortage. The number of doctoral degree holders has not kept pace with the increasing demands from colleges and universities, let alone those from government and planning agencies.

Table 1 shows membership figures for the Association of American Geographers since 1957. Geography has shown a rapid increase within the past few years. The 1957 membership had doubled by 1965, and had more than tripled by 1968, an increase more rapid than most of the other professional associations in the behavioral and social sciences.

Faculty and Enrollment

The average geography faculty in the departments offering doctoral degrees in 1966 included ten full-time equivalent members, approximately one-half the size of other behavioral and social sciences departments (Table 2). In Table 2 the heading en-

TABLE 2 FULL-TIME EQUIVALENT FACULTY, GEOGRAPHY AND OTHER BEHAVIORAL AND SOCIAL SCIENCE DEPARTMENTS IN PHD-GRANTING UNIVERSITIES, FALL 1961, FALL 1966, FALL 1971, FALL 1976

	Other Behavioral and Social Sciences			Geography		
Full-Time Equivalent Faculty, by Year	Total	Department Mean	Per Cent Increase of Total	Total	Department Mean	Per Cent Increase of Total
Fall 1961	8,379	13.6	—	338	7.5	—
Fall 1966	12,044	19.5	44%	455	10.1	35%
Fall 1971	16,628	27.0	38%	666	14.8	46%
Fall 1976	20,472	33.2	70%[a]	836	18.6	84%[a]

Source: Departmental Questionnaire.
[a] Per cent increase from 1966.

titled "Other Behavioral and Social Sciences" includes anthropology, economics (including agricultural economics), history, political science, psychology (including educational psychology), and sociology (including rural sociology). University geography faculty in these departments has increased by 35% since 1961 and lags somewhat behind the increase in other behavioral and social science faculty (44%). The greatest contrast occurred at the assistant-professor level where the other behavioral and social sciences increased nearly three times as much as did geography. Estimates for 1976 show an anticipated geography increase of 84% to an average departmental size of more than 18 as compared with the anticipated average increase of 70% for the other behavioral and social sciences to an average-size faculty of 33. The current manpower shortage is evident in the high figures for unfilled faculty positions (Table 3). As shown in Table 3, the geography departments surveyed had a higher

TABLE 3 UNFILLED FACULTY POSITIONS, BEHAVIORAL AND SOCIAL SCIENCE DEPARTMENTS IN PHD-GRANTING UNIVERSITIES, FALL 1967, BY FIELD

Field	Full-Time Equivalent Faculty (1)	Unfilled Positions (2)	Ratio (2)/(1)
Geography	455	57	.125
Sociology	1,386	113	.082
Anthropology	674	44	.065
Psychology	2,534	144	.057
Economics	1,953	104	.053
Political Science	1,778	90	.051
History	2,612	96	.037

Source: Departmental Questionnaire (inflated).

ratio of unfilled faculty positions to total faculty positions in 1966 than did any of the other behavioral and social science fields. The unfilled geography positions amounted to nearly 13% of the 1966 full-time equivalent faculty as compared with 8% for sociology and approximately 5% for the others.

Enrollment figures provide further evidence of growth of geography in universities. As shown in Table 4, total geography enrollment according to Schwendeman's *Directory of College Geography* went from less than 250,000 in 1956–57 to more than 600,000 in 1967–68, so that the 1957-based index reached 258 in 1967–68.

TABLE 4 TOTAL GEOGRAPHY ENROLLMENT IN COLLEGES AND UNIVERSITIES 1957, 1958, 1967, 1968

Year	Total Enrollment	Number of Geography Departments	Number of Graduate Programs
1956–1957	245,968	213	73
1957–1958	274,136	217	73
1966–1967	581,411	298	115
1967–1968	635,181	351	120

Source: *Directory of College Geography in the United States,* Association of American Geographers, Southeastern Division, ed. J. R. Schwendeman (Richmond: Eastern Kentucky University, 1957, 1958, 1967, 1968).

Degrees and Graduate Enrollment

The number of bachelor's degrees awarded in geography has been relatively high and the number of graduate degrees has been relatively low as measured in terms of growth in the other behavioral and social sciences since 1957. Table 5 shows that since

TABLE 5 DEGREES AWARDED IN 1957 AND 1967 AND PROJECTED FOR 1977—GEOGRAPHY AND TOTAL BEHAVIORAL AND SOCIAL SCIENCES—UNITED STATES AND OUTLYING AREAS, BY LEVEL OF DEGREE

Level of Degree	1957 [a]	1967 [a]	1977 [b]
Bachelor's			
Geography	699	2,163	5,544
Total	49,154	124,595	318,800
Master's			
Geography	182	463	1,232
Total	5,796	18,725	57,400
Doctor's			
Geography	47	79	91
Total	1,677	3,915	8,900

[a] Source: Actual values for 1957 and 1967 obtained from the Office of Education.
[b] Source: Behavioral and Social Science Survey Committee Projections. For a description of the projection methods used see *The Behavioral and Social Sciences: Outlook and Needs* (Englewood Cliffs, N.J.: Prentice-Hall, Inc., 1969), Appendix D.

1957, geography has increased its bachelor's degree production at a somewhat more rapid pace than the over-all rate of increase in the behavioral and social sciences. Geography has, however, lagged behind other behavioral and social science fields in the production of graduate degrees, particularly the doctor's degree. Graduate-enrollment figures (Table 6) suggest a recent growth that is con-

TABLE 6 FULL-TIME GRADUATE ENROLLMENT, GEOGRAPHY AND OTHER BEHAVIORAL AND SOCIAL SCIENCE DEPARTMENTS IN PHD-GRANTING UNIVERSITIES: FALL 1961, FALL 1965, BY LEVEL OF ENROLLMENT; 1967, ALL LEVELS

Graduate Enrollment	Other Behavioral and Social Sciences			Geography		
	Total	Department Mean	Per Cent Increase of Total	Total	Department Mean	Per Cent Increase of Total
All levels						
Fall 1961 [a]	19,059	30.89	–	597	13.21	–
Fall 1965 [a]	30,683	49.73	61	958	21.17	61
First year						
Fall 1961 [a]	8,903	14.43	–	292	6.44	–
Fall 1965 [a]	14,611	23.68	64	492	10.86	68
Intermed.						
Fall 1961 [a]	8,848	14.34	–	266	5.88	–
Fall 1965 [a]	14,284	23.15	61	429	9.49	61
Terminal						
Fall 1961 [a]	1,302	2.11	–	40	0.88	–
Fall 1965 [a]	1,771	2.87	36	38	0.83	–5
New Graduate Students enrolled—1967 [b]	19,213	31.14	31[c]	689	15.27	40[c]

[a] Source: Office of Education data for the universities in the Survey.
[b] Source: Departmental Questionnaire.
[c] Per cent increases calculated from full-time first-year students, Fall, 1965.

sistent with degree trends. Full-time graduate enrollment among the geography departments in the Survey increased at approximately the same rate as the average for the other behavioral and social science departments, but there were markedly greater increases among new, first-year, and intermediate students than among terminal students.

Although the number of doctoral degrees awarded is lagging, there is no consistent evidence available on the magnitude of the shortage. Table 7 shows three different estimates of doctoral degrees expected in geography in 1972 and 1977. The projections by the Survey Committee staff are compared with estimates made by geography chairmen answering the Departmental Questionnaire in winter 1968 and by geography chairmen separately surveyed by the Association of American Geographers in the spring of 1968. Both geography chairmen polls indicated expectations of a sharp upturn in doctor's degrees during the next few years. Chairmen estimates for 1971–72 are more than triple those projected; geography chairmen surveyed anticipated 427 doctor's degrees in 1976–77 rather than the 91 projected by the Survey Committee. Although it seems probable that department chairmen would tend to overestimate

TABLE 7 THREE INDEPENDENT PROJECTIONS OF DOCTORAL DEGREES IN GEOGRAPHY TO 1972 AND 1977

Year	Survey Projections	Department Chairmen (Survey)	Department Chairmen AAG
1971–72	80	265	278
1976–77	91	427	-

Sources, Column 1: Projections made by Behavioral and Social Sciences Survey Committee. The Survey Committee projections presume a continuation of the rate of growth of actual degrees awarded, 1958–1967. Thus, if a field began to expand rapidly prior to 1967, the projections take it into account. However, if expansion is imminent but not already present, the projections will be under estimates. This may be the case for geography.

Column 2: Estimates of department chairmen made from Departmental Questionnaires.

Column 3: Survey by Association of American Geographers in the spring of 1968. Returns were received from 25 of the 32 departments, which accounted for 265 out of 306 total PhD degrees between 1961 and 1965 as well as several departments which had been authorized to offer the doctoral degree since that time. Results are expanded so as to include all doctoral degree-granting departments.

doctoral-degree completions in any given year, the chairmen's estimates exceeded the Survey Committee's projections by only 18% for all the behavioral and social sciences. The magnitude of the discrepancy is great enough, therefore, to suggest that the production of doctoral degrees in geography will probably exceed the Survey Committee's projections. The present and proposed entry of several new departments into doctoral programs should also increase the future supply of doctor's degrees in geography. Current plans indicate that approximately four new departments per year plan to undertake such programs during the next few years.

Manpower Shortage

The net result of the combined trends in the availability of geographers with doctor's degrees and the need for them in universities, colleges, government, and planning agencies shows a present and continuing manpower shortage. One projection, based on the anticipated degree production and on the assumption that all college and university positions should be filled by those holding the doctorate, estimated a shortage of 250 persons in 1966–67 that would grow to nearly 700 by 1975–76. The shortage has been acute in geography for the past few years. In 1968 the Commission on College Geography issued a special report on manpower in American geography. Questionnaire returns indicated that, in the fall of 1966, geography departments added 170 new staff members and replaced 120 retired or departmental members but were unable to fill 145 vacancies. Despite these 145 unfilled vacancies, the department chairmen surveyed indicated that they expected to appoint around 390 new staff members in 1967 or 1968.

The A.A.G. survey shows a strong recent tendency for doctoral candidates who have completed all requirements but the dissertation to accept teaching or government positions. One major department with a current resident graduate enrollment of 56 doctoral candidates reported an additional 32 candidates who had completed all requirements but their dissertations; another with a resident enrollment of 61 reported 20. The Commission on College Geography report included results of a questionnaire sent to a list of geographers who had accepted full-time teaching positions before completing their dissertations. As might be expected, the respondents indicated financial need as the chief reason for leaving graduate school and teaching obligations as the chief obstacle to completing their dissertations. This problem is of particular significance to the manpower shortage in geography because it may be self-accelerating. Not only are doctoral departments, both old and new, handicapped in efforts to expand their own graduate training activities, but the tendency to hire promising candidates before completion of dissertation becomes stronger and thereby reduces the level of doctoral degree production in any given year. The promising graduate student stands to forego a substantial income if he *refuses* a post before completing his dissertation. One possible step toward alleviating the shortage, at least in the short run, would be a large increase in federally sponsored terminal-year or dissertation fellowships. These fellowship funds would be invested in mature and proven graduate students, and would yield a quick return in completed doctorates. Other possible measures include reducing the teaching loads or even subsidizing the student for a time in order for him to finish the degree.

One final facet of the future manpower situation in geography is

TABLE 8 DOCTORAL DEGREES AWARDED IN GEOGRAPHY, 1967; ESTIMATED FOR 1972 AND 1977, BY PRESTIGE RANKING

Number of Departments	Prestige Ranking	1967		1972		1977	
		No.	Per Cent	No.	Per Cent	No.	Per Cent
11[a]	Distinguished and strong	46[a]	58	100	38	143	33
34[a]	Other	33[a]	42	165	62	284	67
Total							
45		79	100	265	100	427	100

[a] Source: Departmental Questionnaire (estimated). Categories based on those established in Allan M. Cartter, *An Assessment of Quality in Graduate Education* (Washington, D.C.: American Council on Education, 1966).

illustrated by Table 8. The departments are divided into distinguished and strong departments, and all others included in the

Survey. The categories used are based on those established by Allan M. Cartter in the report *An Assessment of Quality in Graduate Education,* published by the American Council on Education in 1966. In 1967, the distinguished and strong departments produced 58% of the doctoral degrees; by 1977, they will be producing only 33%. Therefore, there will be a shortage of quality manpower. Those departments currently experiencing the greatest difficulty in obtaining adequate faculty are going to be playing an increasingly prominent role in the training of doctoral candidates.

RESEARCH

Although research activity in geography has shown a relatively rapid increase during the 1960s, its initial base was small, and the average research funding per faculty member was well below the behavioral and social science average in 1967. In addition to the usual problems faced by social science research, there are two problems of particular concern to geographers: support for foreign-area research and the need for more precisely located data.

Research Trends

Table 9 shows the pattern of increase in departmental research funds for geography from nonuniversity sources as well as for the total behavioral and social sciences. The funds for geography more than doubled from 1962 to 1967, as did the behavioral and social science totals. At the time of the Survey, departmental chairmen in geography estimated that research funds would increase 350% by 1972 as compared with the anticipated total increase of 270%.

Although research funds for geography from nonuniversity sources showed a sharp increase from 1962 to 1967, the amount per capita was still below average. Table 10 lists the nonuniversity-department-research funds per full-time equivalent faculty member for 1967. Geography received $2,500 per faculty member as compared to an average of $4,012 for all the behavioral and social sciences.

The expanded research activity as shown by the recent and an-

TABLE 9 DEPARTMENTAL RESEARCH FUNDS FROM NON-UNIVERSITY SOURCES, GEOGRAPHY AND TOTAL BEHAVIORAL AND SOCIAL SCIENCE DEPARTMENTS IN PHD-GRANTING UNIVERSITIES, FISCAL YEARS 1962, 1972, AND 1977

Fiscal Year	Geography	Total Behavioral and Social Sciences
1962		
Amount[a]	$ 438,000	$ 1,140,000
1967		
Amount[b]	$1,140,000	$ 52,510,000
Per cent of 1962	260%	256%
1972		
Amount (est.)[c]	$4,000,000	$142,500,000
Per cent of 1967	350%	270%
1977		
Amount (est.)[c]	$6,000,000	$210,000,000
Per cent of 1972	150%	150%

[a] Source: Derived from the administration report of Federal Grants and Contracts for Research, for 1962, as ratio of such grants for 1967, corrected for missing replies.

[b] Source: Departmental Questionnaire report of nonuniversity sources.

[c] Source: Departmental Questionnaire, needed funds for research.

TABLE 10 DEPARTMENTAL RESEARCH FUNDS FROM NON-UNIVERSITY SOURCES PER FULL-TIME EQUIVALENT FACULTY MEMBER, BEHAVIORAL AND SOCIAL SCIENCE DEPARTMENTS IN PHD-GRANTING UNIVERSITIES, FY 1967, BY FIELD

Department	Research Funds per FTE
Psychology	$9,271
Sociology	4,394
Anthropology	3,858
Economics	3,154
Geography	2,500
Political science	1,558
History	494
Average, all behavioral and social sciences	4,012

Source: Departmental Questionnaire.

ticipated growth rates has been evident in other ways as well. There has been a substantially higher output in recent years of important monographs in various subfields and the cumulative impact of these on the profession and on related fields is only beginning to be felt. A monograph series has been established by the Association of American Geographers. The volume of journal articles has mounted steadily. The *Annals* of the Association of American Geographers has steadily increased its size, and further increases are contemplated. *The Geographical Review* and *Economic Geography*, both leading journals, have large backlogs of scholarly articles, and a new U.S. journal, *Geographical Analysis*, has just appeared and will focus on work in locational analysis. There is also a considerable amount of publication by geographers in journals of allied fields and in foreign journals. More than 1,500 geographic serials are published in 70 countries, 250 of which either publish articles in English or make supplemental use of English. The rapid growth of discussion-papers series issued by many of the doctoral-degree-granting departments is another sign of the mounting research volume.

Sources

Geographic research has received support from many and diverse sources. In a survey conducted by the Association of American Geographers that covered over five hundred research grants made to A.A.G. members between 1964 and 1968, the list included the National Science Foundation and Fulbright-Hays programs, and 25 other U.S. government sources, ten Canadian government agencies, 30 different foundations, 11 private corporations, and a variety of state and local governments. Of the 524 grants in the sample, approximately 40% were from universities, 30% from the U.S. government, and 20% from foundations. When dollar amounts were considered, however, the university grants dropped sharply in importance. Of the U.S. government departments, the Department of Defense was clearly the largest source, particularly the Army Corps of Engineers, the Office of Naval Research, and the U.S. Army Research Office. Other agencies fairly strongly represented were the National Aeronautics and Space Administration, the Department of Transportation, the Office of Economic Opportunity, the Environmental Sciences Service Administration, and the Office of Education. Foundations represented were Ford and Guggenheim, and there was a scattering of grants from the NAS-NRC Foreign Field Program, and the Social Science Research Council Foreign Area Fellowship Program.

Table 11 provides a closer look at the participation of geography

TABLE 11 NATIONAL SCIENCE FOUNDATION RESEARCH PROJECTS, SOCIAL SCIENCE DIVISION—FISCAL YEARS 1966–1968

	1966		1967		1968	
Field	**Number**	**Amount**	**Number**	**Amount**	**Number**	**Amount**
Anthropology	125	$ 3,981,890	142	$ 3,662,850	130	$ 3,608,630
Economics	51	2,345,210	75	3,208,475	66	3,688,950
Geography	5	215,300	15	411,000	22	595,300
Sociology and social psychology	83	3,658,600	107	4,062,700	105	4,038,700
Political science	17	335,650	32	804,800	38	788,098
History and philosophy of science	49	1,023,000	42	806,650	47	829,000
Special projects	15	1,032,370	12	1,957,346	18	1,860,700
Total	345	12,592,020	425	14,913,821	426	15,406,370

Sources: National Science Foundation, Social Science Division.

in the research program of the National Science Foundation, Division of Social Sciences. This table lists research projects only, and does not include faculty fellowships or other NSF forms of aid. Nor does it include support of earth sciences and other environmental-sciences research. Geography's participation in the Social Science program in 1966 was relatively limited, even if one takes into account the smaller number of doctoral candidates in geography. In this case too, however, there is evidence of a recent increase in research activity. From 1966 to 1968, geography has shown a consistent increase in the number and amount of National Science Foundation research grants. The key to a further increase in participation by geographers in NSF research programs seems to lie simply in more geographic research and a greater number of proposals submitted, because the proportion of proposals rejected does not seem to vary significantly from geography to other fields.

Very little support for geographic research has been provided by the Department of State and the Department of Housing and Urban Development, two government agencies that might be expected to have particular interests in contemporary geographic research. A greater concern by these and other agencies for basic social science research, as recommended in the report of the Advisory Committee on Government Programs in the Behavioral Sciences, *The Behavioral Sciences and the Federal Government,* would involve geographers and others in research of ultimate relevance to the problems faced by the federal government.

Foreign-Area Research

Foreign-area research presents a particular problem in geographic study. Approximately 40% of geography dissertations listed in the early 1960s dealt with foreign areas: approximately 60% of the research proposals submitted by geographers to the National Science Foundation, Division of Social Sciences, involve foreign areas. The need for continuing cross-cultural study becomes pressing as more effective models for the study of spatial organization are developed. These models are designed for the study of places, and their application to foreign areas can provide helpful insights into problems of urban growth, circulation, and regional development. The U.S. scholar, in turn, learns much about the models themselves by applying them in a different cultural context. Many assumptions, both implicit and explicit, prove to be more limiting than they were in a familiar context, and the models often yield unexpected results.

Graduate students contemplating research careers that involve foreign-area study are faced with several handicaps. The potential foreign-area specialist in geography must not only acquire a competence in at least one of the increasingly complex topical fields such as those described in "Selected Research Directions," but must also become familiar with many of the historical and cultural specifics of the area, and acquire a relatively strong linguistic competence. Besides these requirements the candidate has unusually large dissertation expenses since field work is almost universally required for foreign-area dissertations. In the past several programs have been used by geographers, including those of the Ford Foundation, the Social Science Research Council, and the National Academy of Sciences-National Research Council. The Foreign Field Research Program administered by the Earth Sciences Division of NAS-NRC and funded by the Office of Naval Research has been a particularly effective one in encouraging the development of foreign-area specialists in geography. The program started in 1955 and has made 97 awards that range from $500 to $13,500; the average expenditure was approximately $60,000 per year. The program was well known in geography and probably encouraged many more students toward careers as foreign-area experts than it could actually help, since unsuccessful applicants with strong proposals were often able to obtain support elsewhere. The Office of Naval Research sponsorship ended in 1969 and new sponsorship is being sought. Meanwhile, the potential foreign-area specialist faces serious research-support problems because of a decline in foreign-area support by foundations. In general, this decline in support may be due to the expectation that the International Education Act would provide adequate stimulus to foreign-area research. Failure to implement this Act has left a serious gap in the support of foreign research which, if not remedied, will significantly reduce future U.S. capabilities in international study.

Data Needs

Although geographic data needs are, in general, similar to those of other social sciences, geographers do have a particular need for precisely located information. The common practice of recording data by areas such as counties or census tracts has serious shortcomings as a basis for the study of spatial organization. The need for more effective use of locational coordinates in data storage was indicated in a 1964 report of the Geographic Coding Subcommittee of the Census Advisory Committee of the Association of American Geographers. The primary advantages cited by the subcommittee were greater ease of aggregating and partitioning data, comparing data collected under different systems, computerizing manipulation, and display of data.

An effective system of locational coordinates could provide great flexibility of aggregation of data and thus permit tabulations for areas other than those for which the data were originally stored. For example, at present it is quite difficult to compare the socio-economic data gathered by census tract with data for school districts. With locational coordinates census tract data could be compared not only with school districts but with market areas, police districts, or by distance bands from expressways or from particular parts of the city. Similarly, the comparability of data systems would no longer be dependent on the type of area unit used for collection purposes. Data gathered by city-planning or transportation-survey agencies, which employ disparate area systems, could be readily coded to compare with census data as could air photos or any other localized sensing information.

An effective system of locational coordinates permits the programming of nearly any data for computer display, whether on a printer, a plotter, or a cathode-ray tube at any desired scale. It can delimit local areas of potential violence or different areas proposed for urban renewal, and quickly assemble information on the relevant socio-economic characteristics and yield large quantities of maps of these characteristics. Transformation into maps of uniform population density, purchasing power, etc., is readily possible if the data involved have locational referents.

Some progress has been made toward effective location of data by the Geography Division of the U.S. Bureau of the Census in the preparation of locational codes for census data in the form of latitude-longitude coordinates for 43,000 designated "National Locational Code Areas." The latter are based on centers of census tracts and other areas; these differ in size and shape, however, and are thereby limited in utility. Plans for providing latitude and longitude and other coordinates for block faces in the 1970 Census of Population seem to be more promising. There are also some problems of disclosure and ease of access by interested scholars which prohibit the most telling use of Census data. Further improvement is needed both in the fineness of locational detail and in ease of access if geographers are to apply their newly developing analytical models to some of society's most serious problems.

RESEARCH TRAINING

Geography shares with many of the other behavioral and social science disciplines a relatively long period between baccalaureate and doctoral degrees, weakness in the development of postdoctoral training, and a discrepancy between graduate research activity and initial teaching assignment. Geography differs from the other behavioral and social science disciplines in its greater dependence on teaching assistantships for graduate student support.

Graduate Student Support

The general pattern of graduate student support for the behavioral and social science disciplines is shown in Tables 12A and 12B, which list both the data from the Departmental Question-

TABLE 12A FINANCIAL AID FOR NEWLY ENTERING GRADUATE STUDENTS, GEOGRAPHY AND OTHER BEHAVIORAL AND SOCIAL SCIENCE DEPARTMENTS IN PHD-GRANTING UNIVERSITIES, FALL 1967, BY TYPE OF ASSISTANCE

	Other Behavioral and Social Sciences			Geography		
Type of Assistance	Number Assisted	Dept. Average	Per Cent	Number Assisted	Dept. Average	Per Cent
Fellowships, traineeships, scholarships	6,392	10.36	57%	201	4.45	44 %
Teaching assistants	2,487	4.03	22	214	4.73	47
Research assistants	1,567	2.54	14	28	0.62	6
Other assistants	407	0.66	4	6	0.14	0.2
Other aid	358	0.58	3	11	0.24	3
			100%			100 %

Source: Departmental Questionnaire; sample expanded to 45 geography departments.

TABLE 12B TYPES OF MAJOR SUPPORT OF GRADUATE STUDENTS IN PHD-GRANTING SOCIAL SCIENCE AND PSYCHOLOGY DEPARTMENTS, FALL 1966, BY FIELD (AS PER CENT OF FULL-TIME STUDENTS)

Type of Support	Anthropology	Economics	Geography	Political Science	Psychology	Sociology
Fellowship and traineeship	36%	31%	26%	32%	38%	39%
Teaching assistant	15	19	30	14	19	19
Research assistant	8	12	9	7	16	12
Loan, self-support, and other	41	38	35	47	27	30
	100%	100%	100%	100%	100%	100%

Source: *Graduate Student Support and Manpower Resources in Graduate Science Education, Fall 1965, Fall 1966* (Washington, D.C.: Office of Planning and Policy Studies, National Science Foundation, June, 1968), p. 90.

naire and data from a study of proposals from graduate departments for support from the National Science Foundation Graduate Traineeship Program in fall, 1966. In both sets of figures, geography shows a greater reliance on teaching assistantships than do other disciplines. According to the Departmental Questionnaire (Table 12A), 47% of the new geography students receiving aid in 1967 were being supported by teaching assistantships as compared with an average of 22% of the new students in other disciplines. The NSF figures (Table 12B) show 30% of all geography graduate students (with or without aid) supported by teaching assistantships, while the other disciplines range from 14% to 19%. The NSF figures also show that the difference between geography and the other disciplines in the number of teaching assistantships is markedly greater for first-year than for advanced graduate students.

Although there are some advantages in early exposure to teaching, there are some serious disadvantages in such a strong dependence on the teaching assistantship, particularly as compared to the research assistantship. Properly used, the research assistantship can provide the new graduate student with an invaluable period of apprenticeship to one or more faculty members. The effect of teaching assistantships on research training is also related to the nature of introductory course work in geography, which has been increasing more rapidly than over-all enrollment during the past ten years. In the past, introductory geography courses have not necessarily reflected ongoing geographic research. Many of these courses have simply been surveys of physical or cultural elements or of world regions. The result is that the teaching assistant seldom receives a realistic introduction to the kind of research problems he is being trained to investigate. There are some encouraging signs, however, both in efforts to relate research and teaching assignments more closely and in a tendency to make more use of the research assistantship. The Commission on College Geography has cosponsored several institutes and new course outlines on the introductory course, and has initiated a series of resource papers designed to place contemporary research activity and findings in a context relevant to introductory-course goals. The geography departments in the Survey also anticipate a relatively greater stress on the research assistantship in the future (Table 13). Although the geography chairmen an-

TABLE 13 NUMBER OF GRADUATE STUDENT TEACHING ASSISTANTS AND RESEARCH ASSISTANTS, GEOGRAPHY AND OTHER BEHAVIORAL AND SOCIAL SCIENCE DEPARTMENTS IN PHD-GRANTING UNIVERSITIES, FALL 1961, FALL 1966; ESTIMATED FOR 1971, 1976

Year and Status	Other Behavioral and Social Sciences			Geography		
	Total	Mean per Dept.	Per Cent Increase of Total	Total	Mean per Dept.	Per Cent Increase of Total
Teaching assistants						
Fall 1961	4,350	7.05	–	251	5.56	–
Fall 1966	8,132	13.18	87%	462	10.24	84%
Est. 1971	11,421	18.51	40	702	15.49	52
Est. 1976	14,043	22.76	23	893	19.74	27
Research assistants						
Fall 1961	2,598	4.21	–	40	0.89	–
Fall 1966	4,504	7.30	73	88	1.95	120
Est. 1971	7,299	11.83	62	195	4.31	122
Est. 1976	9,866	15.99	35	322	7.11	65

Source: Departmental Questionnaire.

ticipate that both teaching and research assistantships will increase more rapidly than in other behavioral and social sciences, they expect a percentage increase in research assistants more than double that of teaching assistants for both 1971–72 and 1975–76. However, any move to reduce the number of teaching assistants will be difficult until the manpower shortage has been alleviated.

Length of Training

The period between the baccalaureate and the doctoral degrees in geography is a long one, as it is in most of the behavioral and social science disciplines. Table 14, based on a tabulation of data collected by the Office of Scientific Personnel, National Research Council, shows a median of 9.3 years elapsed between the baccalaureate and the doctoral degree, roughly the same as the figures for sociology, anthropology, history, and political science. The median figure for years actually registered, however, is 5.3 years, and reveals that predoctoral job opportunities retard doctoral degrees in geography as well as in the other behavioral and social sciences. The years elapsed between baccalaureate and doctor's degrees for physics, mathematics, and chemistry are significantly less, and range from 5.7 to 6.7 years. Although no evidence is currently available for geography, evidence from other fields indicates that there has been no significant lengthening of the various doctoral degree periods during the past ten years.

Postdoctoral training is quite common in physics and chemistry, but quite rare in geography and most of the other behavioral and social science disciplines. Effective programs are needed both for training in related fields and for advanced work in geography. Since mathematical locational models, computerized cartography, and remote sensing have developed quite recently, it is probable that a number of new faculty members are beginning their research careers with inadequate training in these vital fields.

Mathematical Training

A particularly difficult aspect of graduate research training is related to training in mathematics. Entering geography stu-

TABLE 14 TIME ELAPSED—BACCALAUREATE TO DOCTORATE 1958–1966: SELECTED FIELDS

Field	Total Population	Time Elapsed (Years)	Registered Time (Years)
Education	18,397	13.6	6.7
Arts and humanities	15,463	9.8	5.7
Sociology	1,760	9.5	5.7
Anthropology, archaeology	740	9.4	5.3
History	3,921	9.4	5.7
Geography	604	9.3	5.3
Political science, public admin., international relations	2,628	9.2	5.1
Economics, econometrics	4,000	8.6	4.9
Agriculture, forestry	4,130	7.6	4.9
Botany, zoology, biology	3,005	7.6	5.4
Earth sciences	2,577	7.6	5.0
Engineering	11,645	7.0	5.0
Mathematics	4,062	6.7	5.1
Physics and astronomy	6,626	6.6	5.5
Chemistry	11,040	5.7	4.7

Data for other fields are from *Doctorate Recipients from United States Universities, 1958–1966*, NAS-NRC Publication No. 1489, Washington, D.C., 1967, prepared by the Office of Scientific Personnel. Tables 1, 14. Data for Geography (1958–1967) were prepared by Walter Bailey, NAS-NRC, Earth Science Division from the Office of Scientific Personnel data.

dents often have weak mathematical backgrounds, a characteristic shared with several other behavioral and social science disciplines because of the selection processes built into undergraduate and even secondary-school curricula. The problem is complicated by the current manpower shortage. The expansion in mathematical work has occurred so recently that it is difficult for graduate departments to obtain enough adequately trained faculty to develop strong programs.

The basic needs are for stronger fundamental mathematical understanding reinforced by work with mathematical models applied to specific research questions. In geography, at least, it is likely that the attainment of better fundamental mathematical understanding will require remedial work for some time to come. Ideally, course work in calculus, probability theory, and linear algebra would be part of an undergraduate major program but this is not generally the case at present, although a recent survey of undergraduate major programs indicated that geography department chairmen rated mathematics and statistics above all other cognate fields recommended for geography majors.

Given the current need for remedial work, there are a number of alternative approaches to the problem, each having advantages and disadvantages. A concentration in the early graduate period on formal mathematical training reinforced by applied work within the discipline has many obvious long-run intellectual advantages. However, delaying the student's contact with research problems runs the risk of his developing a passivity and decline in motivation. An almost opposite approach is an emphasis on applied work designed to develop a research sense and to heighten motivation before the acquisition of formal background. This approach may lead to a "cookbook" view of mathematical and statistical models, however, and the candidate may not take the time to strengthen his background in his later graduate work. Collaborative arrangements have been attempted with disciplinary discussion periods coupled with series of lectures by mathematicians. It is difficult to coordinate content in such arrangements, however, and they tend to become quite dependent on the efforts of a few interested faculty. One possibility is to take greater advantage of programmed learning because of the relatively standard content of some of the basic mathematics courses.

Whatever the solution, it is clear that there must be some optimal divisions of labor between the behavioral and social scientists and the mathematicians. What kinds of models are most appropriately taught within each discipline under what sorts of background assumptions? More study is needed, as are pilot projects geared to the requirements of behavioral and social science disciplines.

The problem of retraining present faculty members is a separate aspect of the broad problem of mathematical training. Postdoctoral fellowships and institutes should be established for such training in view of the difficulties experienced by most faculty members who participate in regular courses at their home institutions during the academic year. Three introductory NSF institutes in the application of quantitative methods in geography for college teachers of geography have been held and a fourth will be given in 1969, but more opportunities for advanced study are needed. Regular leaves or released time designed specifically for such training would also be helpful.

Research Training Expenditures

Tables 15 through 19 show present and projected research needs for space, for computer services, and for research equipment. Geography's projected needs for all three of these seem to be somewhat greater than average.

Geography, psychology, and anthropology departments require more space per faculty member than do the other behavioral and social science departments (Table 15). Cartographic laboratories

TABLE 15 DEPARTMENTAL SPACE: MEAN SQUARE FEET PER DEPARTMENT AND PER FULL-TIME EQUIVALENT FACULTY MEMBER, BEHAVIORAL AND SOCIAL SCIENCE DEPARTMENTS IN PHD-GRANTING UNIVERSITIES, FISCAL YEAR 1968, BY FIELD

Field	Square feet per department	Square feet per FTE
Psychology	26,243	1,238
Anthropology	10,666	808
Geography	7,717	763
Sociology	7,363	418
Economics	6,500	314
Political science	5,187	266
History	5,786	233

Source: Departmental Questionnaire.

TABLE 16 DEPARTMENTAL SPACE: MEAN SQUARE FEET PER DEPARTMENT AND PER FULL-TIME EQUIVALENT FACULTY MEMBER GEOGRAPHY AND OTHER BEHAVIORAL AND SOCIAL SCIENCE DEPARTMENTS IN PHD-GRANTING UNIVERSITIES, FY 1968; ESTIMATED 1971–72, 1976–77

Occupied and Estimated Space, by Fiscal Year	Other Behavioral and Social Sciences		Geography	
	Square Feet per Department	Per FTE	Square Feet per Department	Per FTE
Space occupied 1968	11,006	548	7,717	763
Estimated, 1971–72	19,210	718	12,850	873
Estimated, 1976–77	22,940	690	16,900	910

Source: Departmental Questionnaire.

and workrooms, map storage, and computational facilities are commonly included in geography departments, and the space needs per full-time faculty member are approximately twice that of economics, history, political science, and sociology. Projected space needs for geography (Table 16) are also high and reflect the anticipated expansion of remote sensing and computerized cartography.

TABLE 17 COMPUTER COSTS THROUGH THE COMPUTATION CENTER: MEAN PER DEPARTMENT AND PER FULL-TIME FACULTY MEMBER, BEHAVIORAL AND SOCIAL SCIENCE DEPARTMENTS IN PHD-GRANTING UNIVERSITIES, FY 1967, BY FIELD

Field	Department Mean	Per FTE
Sociology	$8,450	$480
Psychology	9,290	438
Economics	8,050	388
Geography	3,510	347
Political science	5,350	273
Anthropology	1,050	80
History	285	11

Source: Computation Center Questionnaire.

In 1967 geographers made moderate use of computer facilities (Table 17). Geographers averaged $347 in computer costs per faculty member, slightly less than economics and considerably less than sociology or psychology, but more than political science, anthropology, and history. In the future this figure should increase sharply with the expanded use of mathematical models and the continued development of computerized cartography.

The pattern for research equipment investment by fields is similar to that for space needs (Table 18). Psychology had a considerably higher investment in research equipment than any of the other behavioral and social sciences, and anthropology was second, but the average investment of $2,530 in geographic research equipment per faculty member was considerably greater than that of the four remaining fields.

TABLE 18 RESEARCH EQUIPMENT: TOTAL REPLACEMENT VALUE MEAN PER DEPARTMENT AND PER FULL-TIME EQUIVALENT FACULTY MEMBER, BEHAVIORAL AND SOCIAL SCIENCE DEPARTMENTS IN PHD-GRANTING UNIVERSITIES, FISCAL YEAR 1967, BY FIELD

Field	Total	Department Mean	Per FTE
Psychology	$26,531,580	$221,097	$10,429
Anthropology	2,155,870	42,272	3,202
Geography	919,000	25,530	2,530
Sociology	1,603,430	20,298	1,153
Economics	1,356,590	14,432	697
Political science	522,100	6,067	311
History	288,200	2,719	110

Source: Departmental Questionnaire.

Expectations of increased research expenditures in the future are reflected in Table 19. Geography chairmen expect to be making more research expenditures per faculty member during the next five- and ten-year periods than do chairmen in the other behavioral and social sciences. A significant share of these expenditures is associated with the expected development of computerized cartography. In addition to the SYMAP and other programs designed to produce maps (from standard computers, there is need for a growing

TABLE 19 RESEARCH EQUIPMENT: PRESENT VALUE AND ESTIMATED FUTURE COST PER DEPARTMENT AND PER FULL-TIME EQUIVALENT FACULTY MEMBER, GEOGRAPHY AND OTHER BEHAVIORAL AND SOCIAL SCIENCE DEPARTMENTS IN PHD-GRANTING UNIVERSITIES, 1966–67, 1971–72, 1976–77

	Other Behavioral and Social Sciences		Geography	
Replacement Value and Cost, by Year	Department Mean	Per FTE	Department Mean	Per FTE
Replacement value—1966–67	$57,950	$2,890	$25,530	$2,530
Estimated cost—next five years	52,180	1,950	28,970	1,970
Estimated cost—next ten years	98,290	2,960	68,480	3,690

Source: Departmental Questionnaire.

variety of adjunct equipment such as plotters and different sorts of optical display devices. Remote sensing equipment for ground operation, air operation, and computer linkage systems represent another share of the anticipated expenditures. Estimates are particularly difficult for remote sensing. It is conceivable that the given estimates of future research expenditures could be tripled or quadrupled if present experimental work proves successful.

The costs of the development of a full complement of computerized cartographic, remote-sensing, and optical equipment are such, in fact, that it will be necessary to concentrate the effort at a few specialized centers. It would be most economical to establish an institute designed to accelerate advances in computerized cartographic-information systems, cartographic analysis, and graphic display and coordination with alternative remote-sensing systems. These developments would have a strong and lasting impact on the effectiveness of geographic work in all the fields discussed earlier, because they provide more powerful and versatile tools for the analysis of problems of spatial organization.

4
RECOMMENDATIONS

It has been clear in this survey that there are many opportunities for the expansion and improvement of geographic research. If geography is to have a strong and beneficial impact on the constantly changing patterns of spatial organization of American society, it will be necessary to continue this development. The following broad recommendations are proposed:

1. Geographers should be encouraged to participate more fully in interdisciplinary work with other behavioral and social scientists, and in policy-oriented work.
2. Efforts should be made to alleviate the manpower shortage.
3. Centers or institutes for cartographic research and training should be established.
4. Efforts should be made to further develop and apply remote sensing technology and associated data and information systems.
5. Support programs for foreign-area study should be expanded.
6. Special programs should be undertaken to strengthen the mathematical training of geographers.

These recommendations are briefly discussed below. We recognize that the problems referred to are so complex as to render quite unrealistic any attempt by a small diverse group meeting infrequently to formulate specific prescriptive measures. Nonetheless we believe it to be worthwhile, following our surveys and discussions, to identify those lines of development—either within the field of geography or related to it—that we feel should be encouraged.

The Geography Panel recommends that encouragement be given to the trend toward greater participation by geographers in interdisciplinary work with other behavioral and social scientists, and in policy-oriented work.

Although not all geographers are social scientists, the majority of them do deal with problems similar to those dealt with by other social scientists. The studies cited in this report clearly pertain to social science problems. As the geographers doing such work identify themselves more clearly as social scientists, their contribution to a broad range of social science problems will be more readily recognized. The trend toward the introduction of more explicitly behavioral components into geographic research should be encouraged. Geographers should participate in interdisciplinary research centers focused on certain major problems. The establishment of such centers based on interdisciplinary efforts seems preferable to the establishment of centers based on the inevitably overlapping efforts of separate disciplines. Geographers will have particularly valuable contributions to make in dealing with problems of the city, of regional development both foreign and domestic, and of environmental quality.

Geographers should expect to be involved in the work of planning agencies at the local, regional, and national levels. More initiative should be taken by geographers in experimenting with internship programs in agencies such as the Department of Housing and Urban Development, and the Department of State, and agencies at the federal, state, and local levels similarly devoted to problems of a social, economic, and political rather than a military nature. These internships should enable government workers to spend a year or more in advanced study at the universities as well as provide opportunities for graduate students to spend a year with an agency. The programs for graduate students might be developed partly as summer fellowships, partly as long-term cooperative fellowships involving both graduate study and work in the agencies in different years.

The panel recommends that efforts be made to alleviate the manpower shortage.

The evidence gathered in the Survey and the anticipated trends in doctoral-degree production suggests a continuing shortage of fully trained geographers for academic and nonacademic employment in the decade ahead. Terminal-year support is critical because of the large number of geographers whose dissertation work has been affected by their acceptance of predoctoral posts. Expanded government programs of terminal-year and dissertation fellowships at somewhat higher support levels are needed. Universities should take their own steps to facilitate dissertation completion by expanding support and by providing free time and reduced initial teaching loads to faculty who do not yet have the doctoral degree.

There is also an overemphasis on the teaching assistantship as a basis for supporting graduate education in geography. This base should be shifted toward research assistantships. Also associated with the shortage is the weak development of postdoctoral work. Postdoctoral work in certain subfields is so greatly needed, however, that some programs should be developed and their availability made widely known despite the current general manpower shortage. Postdoctoral programs are particularly needed in basic fields such as computerized cartography, locational analysis, and environmental behavior, in which developments have been quite rapid. These programs should be closely articulated, with supportive work in mathematics, economics, and psychology.

Finally, it is possible that, in view of the manpower shortage and the changing nature of the field, a more organized and continuous effort to strengthen graduate education is needed, as was done for the high school and undergraduate curriculum by the High School Geography Project and the Commission on College Geography. The universities with new doctoral programs might profit particularly from the establishment of a group that deals with problems of graduate education.

The panel recommends the establishment of centers or institutes for cartographic research and training.

Cartographic work is so fundamental to geography, and the cost of experimental work in cartography is so large that some coordination of effort is essential. If one or two large centers for specialized research and training were established—attached to universities already strong in this area—they would create significant savings by pooling equipment and they could develop both predoctoral and postdoctoral programs to help alleviate the current manpower shortage. The centers should place particular stress on the widespread interuniversity participation of cartographers and information-systems specialists. They should also facilitate computer linkages and pool expertise and equipment to accelerate and maximize the impact of new developments on geographic research and education. For each center a substantial sum would be required to assure at least a minimum program. The amount will vary, depending on the nature of the program and on the extent of support and special equipment provided by the parent institution, but it seems likely that approximately $500,000 would be needed each year for each center.

The panel recommends that efforts be made to further develop and apply remote-sensing technology and associated data and information systems.

The advent of remote-sensing instruments in aircraft and satellites to observe and monitor earth resources and patterns on a global scale has almost infinitely expanded the horizons for geographic research. Specially equipped and staffed centers are needed for training and research in the use of new instruments, techniques for data-gathering and analysis, and application of results. Annual costs of such centers might range from $500,000 to several million dollars depending on the stage of development of the data-gathering programs.

Strong support should be given to plans for orbiting earth-resource satellites of different types by the National Aeronautics and Space Administration, the Departments of Interior and Agriculture, and the Navy Oceanographic Office. Geographers should participate with these and other agencies in the cooperative gathering of earth-resource information by aircraft and satellite.

Agreements should be made with other countries to provide international sources of localized remote-sensing and air-photo data. Present security restrictions on the scientific use of existing material should be reduced and access to both foreign and domestic imagery should be improved.

Locational coordinates for gathering and storing data should be used in order to make remote sensing data available from a wide range of newly developing information systems that are readily adaptable to machine-processing and print-out techniques. This should be expanded from the Bureau of the Census to other federal agencies and to states and cities where the use of locational coordinates would render large quantities of state and urban data-bank figures comparable.

The panel recommends that support programs for foreign-area study be expanded.

The study of foreign areas is such an essential part of locational analysis, urban study, cultural geography, and environmental perception that it was not felt necessary to identify foreign-area study as a separate research direction. The application of the developing models and concepts to the differing cultural contexts of a foreign area not only leads to a better understanding of these areas but also provides a better insight into the strengths and weaknesses of the models and their underlying assumptions.

The need for access to foreign areas is critical at the predoctoral level. The expenses of foreign-field research and the stringent requirements of combined linguistic, regional, and topical competence make the recruitment of capable foreign-area specialists extremely difficult. Full subsistence, travel, and field-expense aid at the dissertation stage seems essential. Present programs designed for predoctoral foreign field research should be expanded or new programs developed.

In general, funds for the International Education Act should be appropriated and the plans for the proposed multipurpose Center for Educational Cooperation implemented. More provision should also be made for the support of foreign students in U.S. universities in view of the shortage in most parts of the world of competent social scientists trained in urban and regional development. This training would also strengthen and broaden the communication of ideas and research which has already proven so fruitful between U.S., Swedish, and British geographers.

The panel recommends that special programs be undertaken to strengthen the mathematical training of geographers.

More experimental programs for mathematical training in graduate social science education are needed. Among the possibilities are ongoing collaborative programs with applied mathematicians, development of programmed or computer-assisted learning units for basic mathematical understanding accompanied by applications in particular disciplines, support for intensive summer programs for graduate students and faculty, and postdoctoral programs in basic mathematics for scholars at all levels. Within the universities geographers need to cooperate more closely both with other social scientists and with mathematicians to find optimal programs and divisions of labor for training and to design research programs that will result in more useful models of basic spatial processes.

It is the panel's opinion that the next decade will see an increased participation by geographers in the intensified efforts of behavioral and social scientists to deal realistically with the problems of society. Closer liaison between geographers and other disciplines, and deeper mutual understanding are essential to the effectiveness of these efforts. Areas of apparent overlap should be regarded as areas of mutually reinforcing investigation in which emerging policy recommendations will have been critically screened by scholars of differing orientations. The geographer's views and methods will be of particular use in approaching the increasingly serious problems of deteriorating environmental quality and of the changing spatial organization of national, regional, and metropolitan areas.

APPENDIX A

GEOGRAPHY DEPARTMENTS PARTICIPATING IN THE QUESTIONNAIRE SURVEY

Boston University
University of California, Berkeley*
University of California, Davis
University of California, Los Angeles
University of California, Riverside
University of Chicago
University of Cincinnati
Clark University*
University of Colorado*
Columbia University
University of Denver
University of Florida
University of Georgia
University of Hawaii
University of Illinois*
Indiana University
University of Iowa
Johns Hopkins University
University of Kansas
University of Kentucky
Louisiana State University
University of Maryland
University of Michigan
Michigan State University
University of Minnesota
University of Nebraska
University of North Carolina
Northwestern University
Ohio University*
Ohio State University
University of Oklahoma
University of Oregon*
University of Pennsylvania*
Pennsylvania State University
University of Pittsburgh
Rutgers, The State University
Southern Illinois University
University of Southern Mississippi
Syracuse University
University of Tennessee
University of Texas, Austin
University of Washington
Wayne State University*
University of Wisconsin, Madison
University of Wisconsin, Milwaukee

* An asterisk indicates non-response to the questionnaire.

APPENDIX B ACKNOWLEDGMENTS

The Geography Panel would like to express its gratitude to the many geographers who provided help in the preparation of this report. In the initial phases of the work, requests for position papers were sent to some 70 geographers. These papers proved most useful in our deliberations and subsequent preparation of text. Our only regret is that space and time limitations permitted us to use only a fraction of the many stimulating ideas we received. Following is a list of those geographers and others who provided position papers, extended letters or copies of relevant papers:

Homer Aschmann, *University of California, Riverside*
Lawrence A. Brown, *The Ohio State University*
Brian J. L. Berry, *University of Chicago*
Andrew Clark, *University of Wisconsin*
W. A. V. Clark, *University of Wisconsin*
Edgar C. Conkling, *State University of New York at Buffalo*
Kevin Cox, *The Ohio State University*
William L. Garrison, *University of Pittsburgh*
Clarence Glacken, *University of California Berkeley*
Preston E. James, *Syracuse University*
George F. Jenks, *University of Kansas*
Leslie J. King, *The Ohio State University*
Edward Knipe, *East Tennessee State University*
David Lowenthal, *American Geographical Society*
Robert C. Lucas, *U.S. Department of Agriculture, Forest Service*

Donald W. Meinig, *Syracuse University*
Marvin W. Mikesell, *The University of Chicago*
Peter Nash, *University of Rhode Island*
Gunnar Olsson, *University of Michigan*
William D. Pattison, *The University of Chicago*
Robert W. Peplies, *East Tennessee State University*
Allan Pred, *University of California, Berkeley*
Edward T. Price, *University of Oregon*
Gordon E. Reckord, *U.S. Government*
Arthur H. Robinson, *University of Wisconsin*
Thomas Saarinen, *University of Arizona*
Theodore Shabad, *The New York Times* and *The American Geographical Society*
Frederick J. Simoons, *University of Texas*
Robert H. T. Smith, *University of Wisconsin*
Joseph Sonnenfeld, *Texas A & M University*
David E. Sopher, *Syracuse University*
David Sweet, *Battelle Memorial Institute*
Edwin N. Thomas, *University of Illinois, Chicago Circle*
Norman J. W. Thrower, *University of California, Los Angeles*
Edward L. Ullman, *University of Washington*
David Ward, *The University of Wisconsin*
Christian Werner, *Northwestern University*
Gilbert F. White, *University of Chicago*
Julian Wolpert, *The University of Pennsylvania*

The draft material was sent to 75 geographers including all chairmen of PhD-granting departments. We would like to acknowledge our debt to those geographers and others who provided critical comments of the draft material. In addition to those cited in the Preface and in the above list, this includes the following:

John Birch, *University of Leeds*
John Borchert, *University of Minnesota*
Michael Chisholm, *University of Bristol*
Saul B. Cohen, *Clark University*
Roland Fuchs, *University of Hawaii*
Barry J. Garner, *Aarhus University*
Arch Gerlach, *Department of the Interior*

Peter Haggett, *University of Bristol*
Chauncy D. Harris, *The University of Chicago*
Howard Hines, *The National Science Foundation*
John Kolars, *University of Michigan*
Lewis Robinson, *University of British Columbia*

SUGGESTED READING

A relatively thorough summary of developments in a number of geographic subfields up to the early 1950s is provided by *American Geography: Inventory and Prospect,* edited by Preston E. James and Clarence F. Jones, published for the Association of American Geographers (Syracuse, N.Y.: Syracuse University Press, 1954). Perhaps the best-known treatise on the development of geographic thought is "The Nature of Geography," by Richard Hartshorne (*Annals,* Association of American Geographers, XXIX [1939], 173–658. This work, together with a later monograph, *Perspective on the Nature of Geography,* by the same author ("Monograph Series of the Association of American Geographers," No. 1, published for the Association of American Geographers (Skokie, Ill.: Rand McNally & Company, 1959) provides a scholarly account of the evolution of geographic thought from nineteenth-century Germany to the early post–World War II period. *The Science of Geography,* a report of the Ad Hoc Committee on Geography of the Earth Sciences Division National Research Council (Washington, D.C.: National Academy of Sciences–National Research Council, Publication 1277, 1965) surveys research directions in physical geography, cultural geography, location theory, and political geography. *Models in Geography,* ed. Richard J. Chorley and Peter Haggett (London: Methuen & Co. Ltd., 1967) includes essays on recent research directions in a number of subfields.

Books that provide a closer look at investigative methods in geography include: Arthur H. Robinson, *Elements of Cartography* (New York: John Wiley & Sons, Inc., 1960), Leslie J. King, *Statistical Analysis in Geography* (Englewood Cliffs, N.J.: Prentice-Hall, Inc., 1969), and Brian J. L. Berry and Duane Marble, *Spatial Analysis: A Reader in Statistical Geography* (Englewood Cliffs, N.J.: Prentice-Hall, Inc., 1968).

In the field of locational analysis, a series of articles by William L. Garrison, "Spatial Structure of the Economy: I, II, III, *Annals*, Association of American Geographers, XLIX, No. 2 (1959), pp. 232–39; XLIX, No. 4 (1959), 471–82; L, No. 3 (1960), 357–73 surveys a number of models later applied in geographic research. *Locational Analysis in Human Geography* by Peter Haggett (New York: St. Martin's Press, Inc., 1966) provides a useful summary of the rapidly expanding work of the early 1960s. Another significant book is the translation by Allan Pred of Torsten Hagerstrand's classic work on spatial diffusion, *Innovation Diffusion as a Spatial Process* (Chicago: University of Chicago Press, 1967).

Some idea of the scope of work in cultural geography may be obtained from *Readings in Cultural Geography*, ed. Philip Wagner and Marvin Mikesell (Chicago: University of Chicago Press, 1962) and a selection from the writings of Carl O. Sauer, *Land and Life*, ed. John Leighly (Berkeley: University of California Press, 1965). Two recent examples of work representing quite different approaches to cultural geography are: Donald W. Meinig, *The Great Columbia Plain* (Seattle: University of Washington Press, 1968), and Clarence J. Glacken, *Traces on the Rhodian Shore: Nature and Culture in Western Thought from Ancient Times to the End of the Eighteenth Century* (Berkeley: University of California Press, 1967).

Examples of work in urban geography are: Brian J. L. Berry, *Geography of Market Centers and Retail Distribution* (Englewood Cliffs, N.J.: Prentice-Hall, Inc., 1967) and the series of urban studies of the upper Midwest by John Borchert and others (Minneapolis: "Urban Reports," Upper Midwest Economic Study, University of Minnesota, 1961–1964). Some examples of work in environmental perception are to be found in *Papers on Flood Problems*, ed. Gilbert F. White (Chicago: University of Chicago, Department of Geography Research Paper No. 70, 1961).

Journals and other serial publications in geography are listed and briefly described in *International List of Geographic Serials* by Chauncy D. Harris and Jerome D. Fellman (Chicago: University of Chicago Research Paper No. 63, 1960). There were over 1500 serial publications listed in 1960, 250 of which were either wholly or partially in English. The major general U.S. journals are the *Annals,* Association of American Geographers and the *Geographical Review,* published by the American Geographical Society. Two other journals focus on more specialized aspects of geography. These are *Economic Geography,* published by Clark University, which emphasizes economic and urban geography, and *Geographical Analysis,* published by the Ohio State University Press, which emphasizes theoretical work and the use of mathematical models. The American Geographical Society publishes *Soviet Geography: Review and Translation* to make available in English reports of current Soviet research in geography.

The principal bibliographic source in U.S. geography is *Current Geographical Publications,* issued monthly (except July and August) by the American Geographical Society. It provides a comprehensive listing, with occasional annotations, of recent books and periodical articles, arranged by subject and region. An excellent listing of approximately one thousand books in geography is provided in *A Basic Geographical Library: A Selected and Annotated Book List for American Colleges* (Publication No. 2, Association of American Geographers' Commission on College Geography, 1966).

Abstracts of geographic works are available from two British sources. *Geographical Abstracts* (London: Headley Brothers Ltd.) is issued in four volumes: geomorphology; biogeography, climatology, and cartography; economic geography; and social geography. *Progress in Geography* (London: Edward Arnold Ltd.) is a new annual publication that consists of essays outlining recent gains in selected subfields.